The Art of Ceramic Extrusion

Frank Händle

The Art of Ceramic Extrusion

Second Edition

Springer

Frank Händle
Frank Händle Transfer
Mühlacker, Baden-Württemberg, Germany

ISBN 978-3-031-63470-3 ISBN 978-3-031-63471-0 (eBook)
https://doi.org/10.1007/978-3-031-63471-0

This Springer imprint is published by the registered company Springer Nature Switzerland AG
The registered company address is: Gewerbestrasse 11, 6330 Cham, Switzerland

For Juli, Isla, Marlene, Lorenz and Giorgio and Alva

Foreword

Frank Händle looks back on a long career as the managing director of several renowned machine suppliers for the ceramics industry, as a lecturer at German universities, and as the author of over 90 articles and books.

Besides his reference work "Extrusion in Ceramics", which has also been translated into Chinese, the first edition of *The Art of Ceramic Extrusion* was published in 2019. This book is addressed primarily to extrusion practicians and is now available as a second edition with four new chapters.

Experts and partners on the topics of extrusion additives, simulation, additive manufacturing (AM) by extrusion and piston presses have been brought on board to contribute to this edition.

The chapter on piston presses and on extrusion additives are certainly necessary and complementary additions for ceramics extrusion in practice. Simulation of the flow of ceramic bodies in the extruder, pressure head and die has evolved into a relevant "tool" for extrusion since 2019.

The book contribution "AM by Extrusion", a niche technology among the rapidly growing array of AM processes, demonstrates the potential of this technology with reference to three examples from the field.

For all relevant shaping technologies, AM has shown that it can claim its place where—despite continuous further development—long-established processes have reached their limits. Consequently, here, too, AM is making a valuable contribution to making ceramic components available to user industries, meeting their highest requirements.

After "Extrusion in Ceramics", with this second extended edition of *The Art of Ceramic Extrusion*, Frank Händle has authored another standard reference work.

A book for practicians who want or have to know more about ceramic extrusion in order to solve complex problems in their company's operations. It helps them make informed decisions and identify the potential of extrusion as a method for shaping ceramics.

Karin Scharrer
Editor-in-Chief (2001–2024)
Ceramic Forum International
Baden-Baden, Germany

Preface to the Extended Second Edition

In view of the positive reception enjoyed by the first edition of *The Art of Ceramic Extrusion*, the logical step was to compile a second extended edition, without, however, making any revisions to the aspirations and recommendations formulated in the Chap. 1 "How to Use This Book". Essentially, minor errors in the first edition have been eliminated, while four new chapters have been added: Chapter 17 "Piston Extruders in Ceramics"—co-written with Christian Wieder, Managing Director of Haginger Maschinenbau GmbH & Co. KG, the leading manufacturer in this field. Chapter 19 "Additives for Extrusion"—composed by Dr. Rainer Bartusch, who has held several positions of responsibility over many years at Freiberg University of Mining and Technology, in industry, as well as at the Keramik Institut in Meissen. Chapter 20 "Additive Manufacturing by Extrusion" has turned out to be rather extensive on account of the relevance of this new technology. Three different approaches are described. Following a brief introduction, Sect. 20.5.1 details the joint development of ECT KEMA GmbH and TUD Dresden University of Technology concerning the fabrication of components based on hydraulic hardening of cementitious extrudates in layers by means of a mobile extruder. Florian Storch from TU Dresden University of Technology and Martin Simon from ECT KEMA GmbH co-authored this contribution. In Sect. 20.5.2, Claude Schneider, Managing Director of 3D Minerals in Limoges, describes the extrusion process developed by 3D Minerals for the fabrication of high-end ceramic objects. In Sect. 20.5.3, Dr. Frank Clemens reports in detail about developments at EMPA in Dübendorf near Zurich. In the process developed by EMPA, ceramic objects are generated on the basis of thermoplastic extrusion. In Chap. 21 "Simulation in Ceramic Extrusion" co-authored by Marcus Dörner and Christian Schwarz from IANUS Simulation, the basic principles of CFD simulations for ceramic extrusion are explained.Modern simulation methods open up the possibility to simulate the flow patterns in the pressure head and die of an extruder for different conditions, enabling the generation of high-quality extrudates from rheologically challenging ceramic materials.With modern simulation methods, it is possible to supplant the commonly taken trial and error approach. Moreover, on the basis of simulations, optimum pressure heads and dies can be developed, and changes in the rheology of ceramic bodies can be simulated in different scenarios. I hope that with these four new chapters, the second edition of *The Art of Ceramic Extrusion* will prove even more useful for the practical application of ceramic extrusion.

Mühlacker, Germany Frank Händle

Acknowledgements I would like to extend my most sincere thanks to: Dr. Rainer Bartusch, Willi Bender Anita Berdychowska, Michele Colucci, Michael Gerbert, Katrin Göhlert, Jean and Horst Gössele, Prof. Dr. Jochen Kriegesmann, Fritz and Karl and Jakob Laenger, Dietmar and Markus Lutz, Conrad Schmitz, Regina Vogt.

Email Contacts to the Authors

Dr. Rainer Bartusch: rainer.bartusch@mail.de
Dr. Frank Clemens: frank.clemens@empa.ch
Dr. Marius Dörner: m.doerner@ianus-simulation.de
Frank Händle: f.haendle@haendletransfer.de
Dr. Christian Schwarz: c.schwarz@ianus-simulation.de
Martin Simon: msimon@ect-kema.de
Florian Storch: florian.storch@tu-dresden.de
Claude Schneider: claude.schneider@3d-minerals.com
Christoph Wieder: c.wieder@haginger.com

Contents

1 How to Use This Booklet ... 1
1.1 For Whom This Booklet Is Written ... 1
1.2 Some Suggestions for the Use ... 2
1.3 Conclusions ... 3
Bibliography ... 4

2 What Do We Understand by Extrusion in Ceramics? ... 5
2.1 Shaping Methods in the Ceramic Industry ... 5
2.2 What Is Extrusion? ... 7
2.3 Criteria for Extrusion ... 9
2.4 Feeding Systems ... 11
2.5 Auger and Barrel ... 12
2.5.1 Auger ... 13
2.5.2 Barrel ... 13
2.6 De-airing ... 14
2.7 Extruded Ceramic Products ... 15
2.8 Conclusions ... 15
Bibliography ... 15

3 Ceramic Materials—The Malice of the Object ... 17
3.1 Definition of Ceramics ... 17
3.2 Classification of Ceramic Materials ... 18
3.3 Classification of Ceramics According to Applications ... 18
3.4 The Main Properties of Ceramic Bodies for Extrusion ... 20
3.4.1 Plasticity ... 20
3.4.2 Yield Point, Viscosity and What Mr. Bingham Has to Say ... 20
3.4.3 Homogeneity ... 22
3.4.4 Grain Size Distribution ... 22
3.4.5 Wall Friction ... 23
3.5 Conclusions ... 23
Bibliography ... 24

4 A Little Bit of Theory ... 25
4.1 What This Chapter Is About ... 25
4.2 Basic Terminology ... 26
4.3 Scenarios of Pressure Build-up in the Extruder ... 28
4.3.1 Scenario 1: The Notorious "Backflow" ... 28
4.3.2 Scenarios 2 and 3—"Backflow" Again, But for Other Reasons ... 29
4.4 The Benbow–Bridgwater Equation ... 30
4.5 How to Calculate the Output of an Extruder ... 31
4.6 Conclusions ... 33
Bibliography ... 33

5 How to Determine an Extruder and Which Data Do We Need? ... 35
5.1 The Decision Procedure and the Database ... 35
5.2 The Decision Model ... 37
5.3 About the Sense and Nonsense of Questionnaires ... 38
5.4 Conclusions ... 40
Bibliography ... 40

6 Plasticity or the Great Unknown ... 41
6.1 What Does Plasticity Mean in Ceramics? ... 41
6.2 The Two Plasticities ... 41
6.3 How Do We Measure the Two Plasticities? ... 42
6.4 Factors Determining Plasticity of a Ceramic Body ... 44
6.5 Two Methods to Measure the Development of Plasticity ... 45
6.5.1 Good-Natured and Nasty Ceramic Bodies—the Consistency Curve ... 45
6.5.2 Kneading Curve ... 46
6.6 Plasticity or Consistency or Extrudability? ... 47
6.7 Conclusions ... 48
Bibliography ... 48

7 Mixing and Kneading ... 49
7.1 Short Introduction ... 49
7.2 Mixing ... 49
7.3 Kneading ... 51
7.4 Macro-mixing and Micro-mixing ... 53
7.5 Buffer Systems or Dosing Strainers? ... 53
7.6 The Ceramic Body as Dispersed Material System ... 54
7.7 Conclusions ... 54
Bibliography ... 55

8 Extrusion as a System 57
8.1 What Do We Understand by a System? 57
8.2 Components of a System 58
8.3 Consequences 59
8.4 Conclusions 61
Bibliography 62

9 Laminations and Chaos 63
9.1 What Do We Understand by Laminations? 63
9.2 Formation of Laminations 64
9.2.1 The Pioneering Studies by BCRA 65
9.2.2 Where Are Laminations Generated in Extrusion? 66
9.3 Types of Laminations 66
9.3.1 Flow Laminations 67
9.3.2 Cutting Laminations 68
9.3.3 Rotational Laminations 69
9.3.4 Hub Laminations 69
9.3.5 Laminations Caused by Insufficient De-airing 70
9.3.6 Combined Laminations 70
9.4 What Can Be Done? 71
9.5 Conclusions 71
Bibliography 72

10 About the Brave New World of Sensors and Control 73
10.1 Introduction 73
10.2 The Mechatronic Model 74
10.3 Why There Is a Compelling Necessity for Better Measuring Systems for Ceramic Extrusion? 75
10.4 What Are the Most Important Operating Data for Extrusion? 75
10.5 Aggregated Measured Values/Operating Data 77
10.6 Conclusions 78
Bibliography 78

11 Ancillary Equipment and Accessories 79
11.1 The Most Important Accessories for the Extrusion Unit 79
11.2 Foundation, Base Frame and Supporting Structure 80
11.3 Control 80
11.4 Cooling/Chilling of Augers and Barrels 81
11.4.1 Why and When Must an Extruder Be Cooled? 81
11.4.2 Selection of the Cooling Unit 81
11.4.3 The Major Operating Data for the Cooling System 82
11.4.4 Also to Be Noted 83
11.5 Vacuum Pump and Vacuum Monitoring 83
11.5.1 Definition of the Vacuum Pump 83
11.5.2 How to Measure the Level of Vacuum Correctly 84
11.5.3 Types of Vacuum Pumps 84

11.5.4 Regulation and Measurement of the Vacuum 86
11.5.5 Checking and Testing the Vacuum 86
11.6 Hydraulic and Pneumatic Equipment 87
11.7 Screen and Die Changers 87
11.8 Infrastructure 87
11.9 Conclusions 88
Bibliography 88

12 Pressure Heads and Dies 91
12.1 What Matters 91
12.2 Typical Pressure Head/Die Combinations 92
12.2.1 Example 1: Pressure Head/Die Combination to Produce Large Brick Blocks 92
12.2.2 Example 2: Pressure Head/Die Combination to Produce Stiff-Extruded Face Bricks 93
12.2.3 Example 3: Typical Pressure Head/Die Combination to Produce Honeycombs 94
12.3 A Look Over the Fence 95
12.4 Conclusions 96
Bibliography 97

13 Extrusion Problems and How to Solve Them 99
13.1 A Bit of Help for Self-Help 99
13.2 Conclusions 103
Bibliography 103

14 Pelletizing by Extrusion 105
14.1 What Do We Understand by Pelletizing? 105
14.2 Pelletizing by Extrusion 106
14.3 Plasticity as a Prerequisite for Pelletizing 107
14.4 Pellet Geometry 107
14.5 Essential Features of a Pellet Extruder 109
14.6 Dies and Matrix Plate 109
14.7 Matrix Plate Changer 111
14.8 Other Pelletizing Methods 111
14.9 Conclusions 112
Bibliography 112

15 Wear, Tear and No End 113
15.1 Wear—A Real Challenge 113
15.2 Types of Wear 114
15.3 Summary of the Most Important Wear Solutions 115
15.4 Roughness of Auger Surfaces 115
15.5 "Where There Is No Friction There Is No Wear" 118
15.6 Function of the Barrel in Extrusion of Ceramic Materials 119
15.7 Conclusions 120
Bibliography 120

16 The Ceramic Production Laboratory ... 121
16.1 What Are the Reasons for Having a Production Laboratory? ... 121
16.2 The Types of Production Laboratories ... 121
16.3 Frequency of Sample Testing ... 122
16.3.1 When and How Often? ... 122
16.3.2 The Worst Case ... 123
16.4 The Subject of Random Samples ... 123
16.5 A Short Real-Life Story ... 125
16.6 Methods and Instruments ... 126
16.7 Conclusions ... 128
Bibliography ... 129

17 Piston Extruders in Ceramics ... 131
17.1 Types of Piston Extruders ... 131
17.2 Working Principle of a Piston Extruder ... 132
17.3 Advantages of Piston Extruders ... 133
17.4 Components of the Piston Extruder ... 133
17.4.1 Hydraulic System and Hydraulic Piston ... 134
17.4.2 Base Frame and Press Frame of the Piston Extruder ... 135
17.4.3 Material Container ... 136
17.4.4 Die Holder ... 137
17.4.5 Vacuum System ... 138
17.4.6 PLC and Data Logging ... 139
17.5 Auxiliary Equipment ... 140
17.5.1 Handling and Cutting Device ... 140
17.5.2 Automation of a Piston Extruder ... 141
17.6 Further Developments ... 142
17.7 Conclusions ... 143
Bibliography ... 144

18 Does that Exist—The Optimal Extruder? ... 145
18.1 A Sad Story ... 145
18.2 Questions and Criteria ... 146
18.3 Optimal Extruder = Optimal Extrusion? ... 147
18.4 Conclusions ... 148
Bibliography ... 148

19 Additives for Extrusion ... 149
19.1 Objectives for the Use of Additives ... 149
19.2 Types of Additives ... 150
19.2.1 Plasticizers ... 150
19.2.2 Lubricants ... 154
19.2.3 Additives to Influence the pH Value ... 157
19.3 Technological Aspects of the Use of Additives ... 157
19.4 Conclusions ... 159
Bibliography ... 159

20 Additive Manufacturing by Extrusion in Ceramics 161
20.1 Introduction 161
20.2 Theoretical Principles of Extrusion 162
20.3 Configurations 162
20.4 Hardening of the Green Extrudates 163
20.5 AM by Extrusion—examples in Development and Practice 163
20.5.1 Development by ECT-KEMA GmbH/TU Dresden 163
20.5.2 3D Minerals Development 168
20.6 Three Machines, Three Sizes 173
20.6.1 AM by Thermoplastic Extrusion 174
20.7 Prospects 179
20.8 Conclusions 180
Bibliography 181

21 Simulation in Ceramic Extrusion 183
21.1 Introduction 183
21.2 The Principle of CFD Simulations 184
21.3 Determining Model Parameters and Boundary Conditions 187
21.4 Error Estimation 189
21.4.1 Programming Error 189
21.4.2 Iteration Error 189
21.4.3 Discretization Error 190
21.4.4 Modelling Error 191
21.5 Application of Simulation Methodology 191
21.5.1 Ceramic Masses in Battery Systems 192
21.5.2 Extrusion of Profile Geometries 193
21.5.3 Calculation of Extrusion Screws 194
21.6 Conclusion 194
Bibliography 194

22 A Look-Ahead 197
22.1 Predictions, Strategies and Wishes 197
22.2 Gaps and Blind Spots 198
22.2.1 The Know-How Gap 198
22.2.2 The Wear Protection Gap 199
22.2.3 The Sensor Gap 200
22.3 The Contribution of Artificial Intelligence 201
22.4 En Route to the "Smart Extrusion Concept" (SEC) 201
Bibliography 202

Chapter 1
How to Use This Booklet

This book of mine has little need of preface, for indeed it is all preface from beginning to end.
—D'Arcy Wentworth Thompson

Abstract For whom this book is intended. What the reader can and cannot expect. How the reader should utilize it and a list of the basic literature.

Keywords Extrusion · Ceramics · Literature · Questionnaire · Target group

1.1 For Whom This Booklet Is Written

In 2004, I had the privilege to act as editor for the book "Extrusion in Ceramics". In the intervening years, "Extrusion in Ceramics" has been translated also into Chinese and has become established as standard literature on this topic. This current booklet "The Art of Extrusion" is neither an abridged version of "Extrusion in Ceramics", nor is it aimed at the same target group, although there are many content-related overlaps. The title "The Art of Ceramic Extrusion" has deliberately been chosen because in ceramic extrusion manual and management skills are needed alongside theoretical knowledge. In this context, the word "art" stands for "craftsmanship". In Germany, if someone refers to another person by saying "he knows his craftsmanship", this means a high level of professionalism in a targeted implementation of knowledge and experience.

For a manager of a ceramic extrusion plant, knowledge of all the tools of the Toyota Production System (TPS) simply does not suffice. And it is not enough for a recently graduated brilliant material scientist involved in the production of roofing tiles to handle ceramic phase diagrams, whilst not understanding anything about ceramic and mechanical process engineering. And finally, it does not help the self-assured practitioners to boycott the opening of new avenues with the opinion that similar action had not worked some 30 years ago.

F. Händle, *The Art of Ceramic Extrusion*, https://doi.org/10.1007/978-3-031-63471-0_1

It is for this heterogeneous target group of people, who in the past were either rarely or only in a very one-sided manner faced with the extrusion of ceramic materials, that I have set out the "basics".

The craft of ceramics is one of the oldest trades in the history of human civilization. It covers the most diverse of sectors, such as pottery, the production of a variety of bricks and tiles, filigree honeycombs, sanitary ware, tableware, fuel cells or insulators. Not only the most varied ceramic materials, but also basically different shaping methods are applied—see Chap. 2.

Here, we concentrate on the shaping method of extrusion as employed both in conventional, as well as technical, ceramics. Although I am well aware of the fact that the problems associated with the extrusion of facing bricks differ in many details from those problems encountered with the extrusion of honeycombs or insulators, there does exist a major intersection in terms of common and extrusion-typical factors and problems.

Extrusion as a shaping method is widely used in the plastics industry, predominantly the thermoplastic process, i.e. melting of the plastic feedstock at high temperatures during extrusion. In addition, extrusion is adopted in many sectors as a preparation method for the activation of compounds, for dispersing, de-watering or pelletizing—see Chap. 14. Can any of the know-how and techniques developed in these sectors be transferred to the extrusion of ceramic bodies? In the majority of cases, only to a limited extent due to the very different rheological parameters or consistency of the materials to be processed. In the coming years, we will however be able to most certainly benefit from the sophisticated simulation methods developed for plastics extrusion: the first tests look very promising.

Being an enthusiastic supporter of the system-theoretical approach, I would point out that extrusion, being a subprocess within a production process consisting of upstream or downstream subprocesses, means that there is a need to know the qualitative and quantitative dependencies and feedbacks between the individual subprocesses. More information can be found in Chap. 8.

1.2 Some Suggestions for the Use

The library of the Technische Universität in Berlin was once one of my favourite places to be—apart from the Berliner Philharmonie, the Bertold Brecht Theatre in former East Berlin, and a pub called "Zwiebelfisch" at the Savigny Platz—where I met with friends. In the library of the TU, I had to wait sometimes for months for foreign books and articles for a bibliography on "Management in R&D" or for my thesis. And how are things today in the brave new digital world? I do not believe that it is really necessary to describe this to you. The convenient and fast availability of any kind of information is the reason I keep referring to certain keywords in Wikipedia or on the Internet. Some of the contributions in Wikipedia are excellent; unfortunately, this cannot be said for the contributions on extrusion. For specific terms such as dissipation or Bingham fluid or more comprehensive terms like "problem-solving" or "catalysis", please refer in case of need to relevant sources on the Internet.

Now and again, I refer to the "ECT Newsletter" which was periodically circulated worldwide to around a thousand interested parties between 2003 and 2013 and which had a great response. This is likely due to the fact that the newsletter represented an attempt to transfer practice-oriented know-how; as hopefully does this concise book too. I have endeavoured to incorporate several questionnaires and checklists, which should assist in the planning and evaluation of extrusion plants prior to the decision-making process and later after implementation, as well as in analysing problems.

In good operating manuals for extruders, one finds guidance for troubleshooting, only later to frequently discover that unfortunately these hefty documents are gathering dust in some cupboard or other; I would strongly recommend that manuals remain handy.

As I have assumed that many readers may only be interested in individual chapters of this book, it could not be avoided that parts of certain chapters are redundant. Finally, I have attempted to avoid the solemn scientific jargon and confine myself to the essentials in language which are easy to understand. "Much was learnt but little was understood" should not be the result of your reading!

1.3 Conclusions

- May I ask you to first of all study the "content" and the above introduction, so that you are aware of what to expect before going on to further chapters.
- The redundant sections in some chapters and the hypertexts should help to maintain the overall context.
- If you find the questionnaires helpful, please supplement and detail these questionnaires with your individual and specific points.
- I would be very grateful for any feedback, suggestions, ideas and critiques.

Here is a list of books that I consider essential. I have restricted myself to only 11 books in the awareness that some relevant sources are missing. Specific literature is listed at the end of each chapter. In addition, articles of interest can be found in the pertinent magazines and the publications of the different ceramic societies and associations.

A selection:

- Journal of the European Ceramic Society
- Journal of the American Ceramic Society
- Ceramic Forum International
- Ceramics International
- Advances of Applied Ceramics
- International Journal of Applied Ceramic Technology

- Journal of Ceramic Science and Technology
- Journal of the Ceramic Society of Japan
- Journal of Advanced Ceramics
- Keramische Zeitschrift
- Ziegelindustrie International.

etc.

Bibliography

1. W. Bender, F. Händle, *Brick and Tile Making* (Bauverlag GmbH, Wiesbaden und Berlin, 1982). (Already published in 1982 this book contains relevant articles regarding extrusion of ceramics, especially those of C.F. Pels Leusden)
2. W.E. Brownell, *Structural Clay Products* (Springer, New York, 1976)
3. J. Heinrich, C.M. Gomes, in Introduction to ceramics. www.ceramics-processing.com. Accessed 11 Mar 2018 (All the fundamentals regarding the processing of ceramics, easy to obtain. A must!)
4. F. Händle (ed.), *Extrusion in Ceramics* (Springer, Berlin, 2009)
5. W.D. Kingery et al., *Introduction to Ceramics* (Wiley, New York, 2011). (The true classic for material scientists in ceramics)
6. E. Krause et al., *Technologie der Keramik* (VEB Verlag für Bauwesen, Berlin, 1985)
7. G.Y. Onoda, et al. (eds.), *Ceramic Processing Before Firing* (Wiley, New York, 1978). (One of the first books about processing in ceramics, still valuable to read)
8. J.S. Reed, *Principles of Ceramic Processing* (Wiley, New York, 1995). (Another must for everybody dealing with ceramic processing)
9. D.W. Richerson, *Modern Ceramic Engineering* (Taylor & Francis, Roca Raton, 2006)
10. D.W. Richerson, *The Magic of Ceramics* (Wiley, Hoboken, 2012). (Read this wonderful, illustrated book to enjoy the cultural and technological diversity of ceramics)
11. R. Telle (ed.), *Salmang/Scholze Keramik* (Springer, Berlin, 2007). (A treasure trove for material-orientated readers. Unfortunately at this moment only available in German)
12. W. Kollenberg (Hrsg.), *Technische Keramik* (Vulkan Verlag, Essen, 2009)

Chapter 2
What Do We Understand by Extrusion in Ceramics?

Abstract The position of extrusion as a major method of shaping in ceramics is defined and evaluated. The process of extrusion in ceramics and the different types of extrusion units are classified focusing on extrusion by single-screw extruders. The different configurations of screw extruders, the main criteria for ceramic extrusion and the relevant components of an extrusion unit are outlined.

Keywords Ceramic extrusion · Shaping methods · Extrusion units · Auger and barrel · De-airing

2.1 Shaping Methods in the Ceramic Industry

Extrusion is one of the shaping processes used in the ceramic industry, parallel to other methods such as casting, dry pressing, injection moulding and isostatic pressing. Extrusion is always employed when it comes to producing endless profiles, which are then cut to the desired length by a cutting mechanism positioned upstream of the extruder. Only in the special case of extrusion by the indexing method, such as the one used for the production of the so-called calotte-shaped pipes or ceramic vessels, the extrusion and cutting of the products are synchronized in one process (Fig. 2.1).

Extrusion in the ceramics industry is always applied mainly in the manufacture of mass-produced articles, such as bricks, pipes and different profiles of all kinds, including honeycombs featuring wall thicknesses of less than one millimetre. One of the first considerations when manufacturing ceramic articles is to select the right shaping method to suit the end product. In order to assess this, one needs to be acquainted with the different shaping systems. If only a very small and complicated number of items are to be produced annually, the casting process or "additive manufacturing" system might be a better option than extrusion. Should low geometrical tolerances and uncomplicated parts be demanded without post-processing, injection moulding or dry pressing appears to be the better alternative. For example, each of

---

extrudere (Latin) = to bump out, to displace, to drive out.

F. Händle, *The Art of Ceramic Extrusion*, https://doi.org/10.1007/978-3-031-63471-0_2

Fig. 2.1 Shaping methods in ceramics

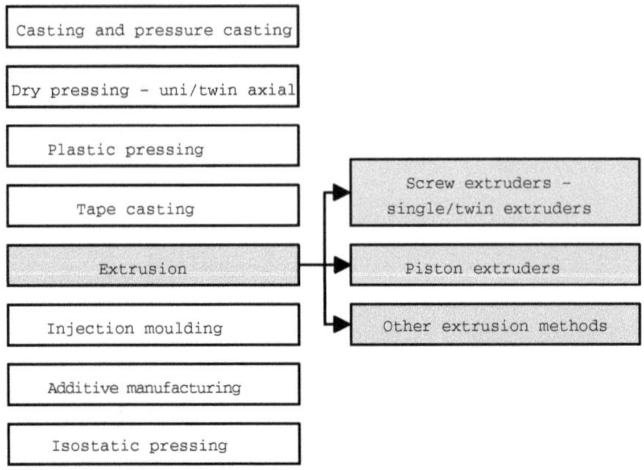

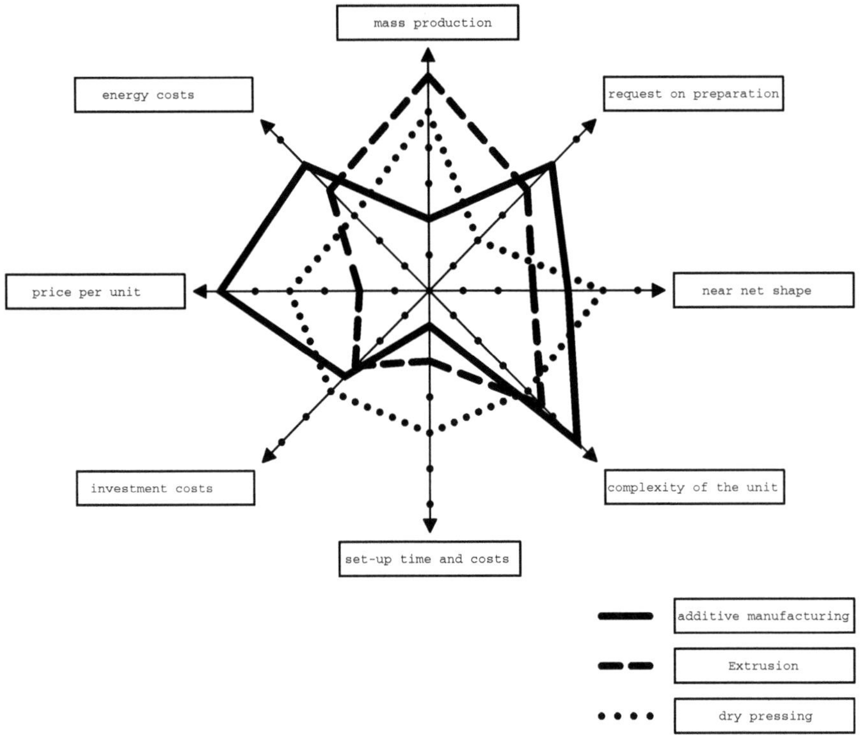

Fig. 2.2 Connotative level to compare shaping methods in ceramics

these shaping processes has its pros and cons which make it better, or less suited, for the product in question. The criteria needed to assess the process to be adopted must be established on a case-to-case basis, particularly since technical progress and market demands bring shifts both in terms of the individual processes and also in respect of the factors to be taken into consideration. A suitable tool for illustrating the comparison of the different process methods and several criteria is presentation on a so-called connotative level.

Figure 2.2 shows an example covering three shaping methods, whereby the chosen criteria can be replaced or supplemented by other and further criteria.

2.2 What Is Extrusion?

What could a child see when it looks at an extrusion unit in a brick factory or in a plant for the production of honeycombs? It would see a machine that obviously consists of three parts: a so-called pug sealer, a bath and extruder below connected with a kind of chamber—Fig. 2.3.

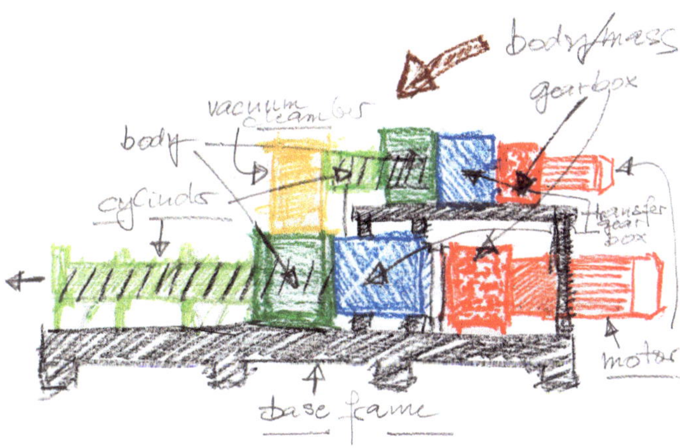

Fig. 2.3 Extrusion unit through the eyes of a child

Let us assume you worked as a taxi driver, an investment banker or a jazz drummer, financial mathematician, or you have developed apps that no one needs. One fine day you find a job in a brick factory, or in a company producing honeycombs. The next morning you are to be familiarized with the extruder at your new workplace. Never heard of extrusion? You seek help from Wikipedia, but soon discover that there are none or hardly any contributions on the subject of ceramic extrusion, despite the fact that thousands of extruders are employed as shaping units in the ceramics industry. Unfortunately, the authors of these contributions are obviously referring to thermoplastic extrusion commonly employed for the extrusion of plastic materials. A friend then suggests you look up the relevant and confusing literature, but a brief

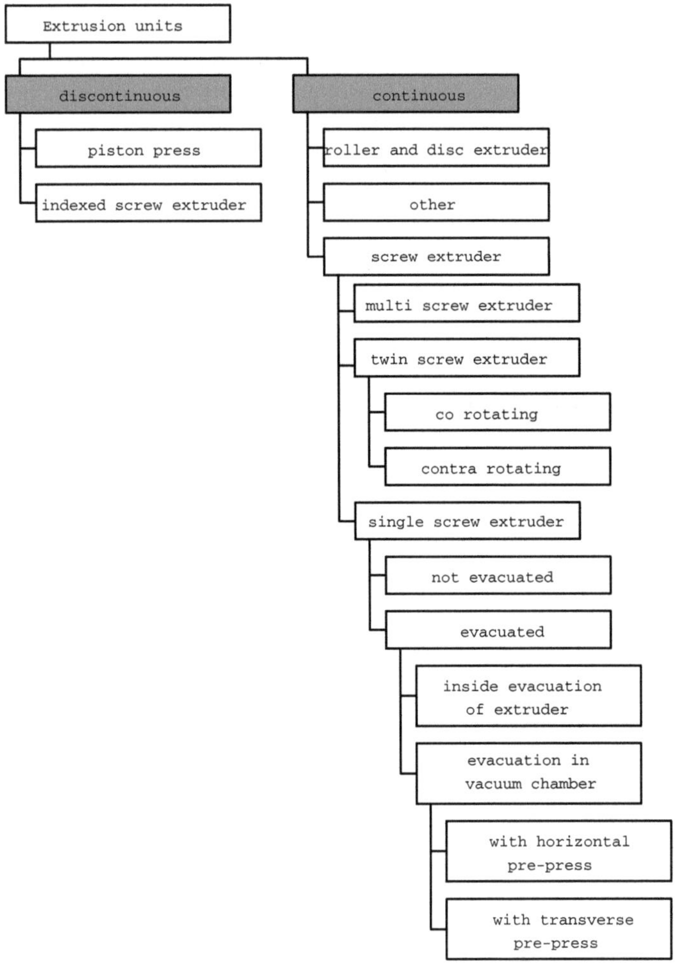

Fig. 2.4 Extrusion units in ceramics

glance is enough to tell you that the information supplied is too theoretical and too extensive. So let us try to give a concise explanation: the terms "extruder" and "extrusion" are derived from the Latin verb *extrudere*, which means to push out. The extrusion process is applied in many fields: in the plastics industry, e.g. for the extrusion of window profiles; in the food industry for the production of pasta, bread and dog food; for the production of wood pellets; or for the extrusion of aluminium profiles, etc. Each time you push toothpaste out of the tube onto your toothbrush in the mornings, you are also using the extrusion system! Today, there are countless applications for extrusion in many different variations and in the most diverse of fields. Here, we concentrate on extrusion in the ceramics industry, particularly on extrusion by single-screw extruders (Fig. 2.4).

2.3 Criteria for Extrusion

Extrusion as a process technology in the ceramic industry can be differentiated according to the following most important criteria:

- By the field of application, e.g. in the building industry, fine ceramics industry, engineering ceramics, refractory industry.
- By the extrusion temperature.
- By the de-airing of the body: the difference between de-airing extruders and extruders without de-airing—for more details, check up Chap. 4.
- By the configuration of the primary pug sealer and extruder, either in transversal arrangement or in cascade configuration.
- By the pressure in either low-pressure, medium-pressure, high-pressure or maximum-pressure extrusion. In terms of ceramic extrusion, low pressure refers to pressure rates up to 50 bar, medium pressure refers to pressure rates up to 100 bar, high pressure up to 200 bar and maximum pressure is pressure rates up to 500 bar.
- By the extrusion unit employed, e.g. screw extruder used as single or twin-shaft version, piston extruders and other special versions such as roller presses and disc presses.
- By the direction of extrusion, either horizontal or vertical or inclined version.
- By the method of operation—either continuous extrusion, discontinuous extrusion or indexing extrusion.
- By the type of material which comes into contact with the body, e.g. corrosion-resistant stainless steels, wear-resistant coatings for abrasive bodies through to components made of ceramic, plastic or hard metal.
- By the degree of shearing intensity taking place during the shaping process of the ceramic body.
- By the throughput rate: starting from throughput rates of ≤ 1 l/min of compact material up to rates of ≥ 100 m^3/h.
- By the size and geometry of the augers or pistons for piston extruders, ranging from 10 mm for micro-extrusion up to 1600 mm in diameter. In Fig. 2.5 You see a photo of the world's largest vertical extruder for the production of billets for insulators.

Fig. 2.5 Vertical extrusion of billets for insulators

I am certain more overlapping distinguishing features could be found. It is however a common feature of all extruders, irrespective of whether they are used for extruding final or intermediate products, for the extrusion of compact columns, or complex profiles such as honeycombs, whether a single or multiple column is extruded, whether cold or hot extrusion is performed; the extruder always extrudes a virtually endless column which is geometrically defined by the die (Fig. 2.6).

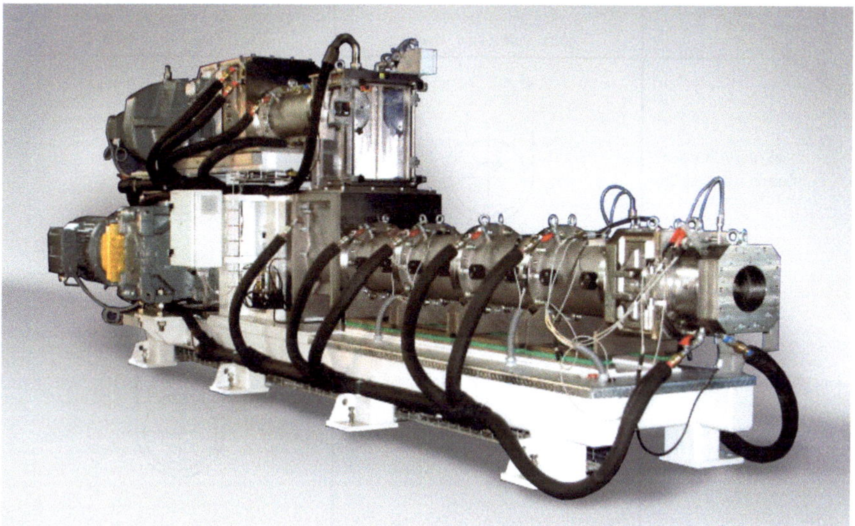

Fig. 2.6 High-tech vacuum unit for advanced ceramics

2.4 Feeding Systems

In practise, the de-airing extrusion units most frequently used consist of a primary pug sealer and the extruder as such, with a vacuum chamber installed in between for de-airing of the ceramic body. The feeding system has a decisive function in ensuring the perfect operation of the system.

Figure 2.7 shows a schematic presentation of the most commonly employed feeding systems. Care must be taken that there is no overfeeding or underfilling of the extruder which may cause various problems in de-airing; in the case of overfeeding, this may result in a "backflow" in the extruder. Consequently, some synchronization is needed between the feeding and the extrusion system which can be supervised and controlled by achieving a suitable feeding level and by using pressure sensors.

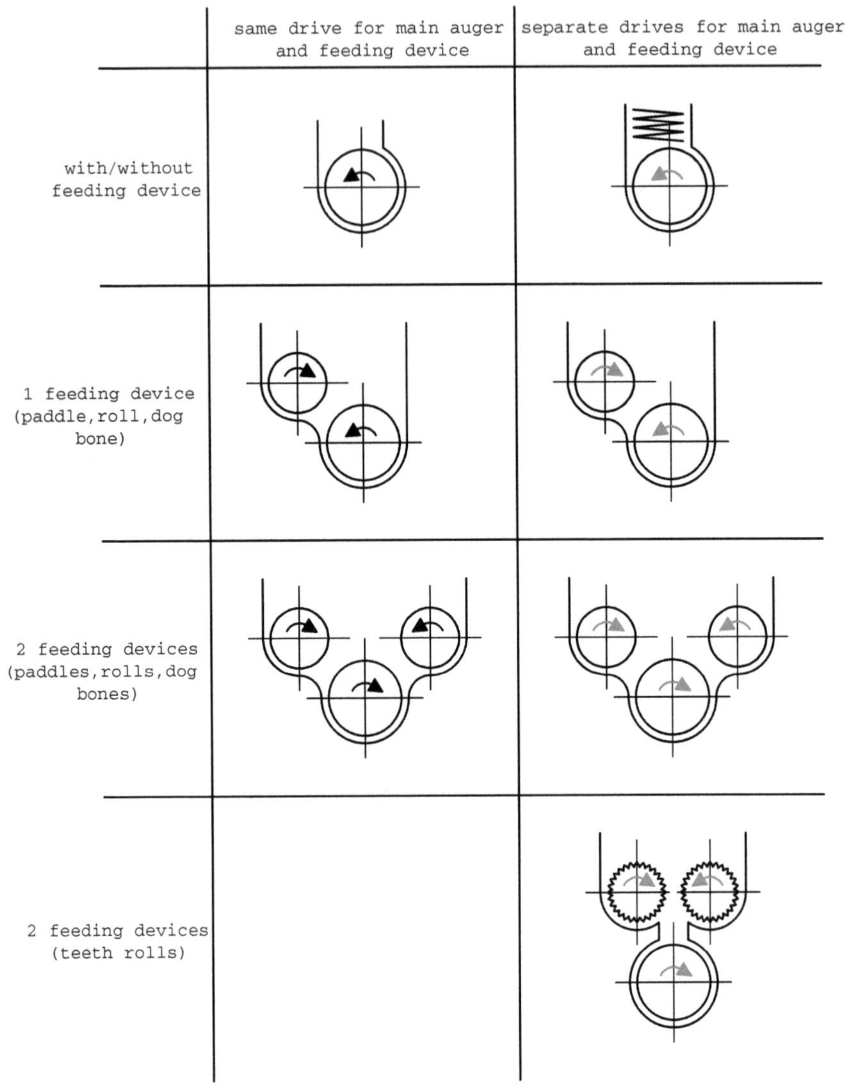

Fig. 2.7 Feeding systems

2.5 Auger and Barrel

This brings us to the two most important elements of an auger extruder.

2.5.1 Auger

Geometric data for auger and barrel is shown in Fig. 2.8:

D_a Outside diameter
D_i Inside diameter
s Pitch
n Thickness of wing
g Gap between auger and cylinder
F_L Leak flow
F_1 Drag flow

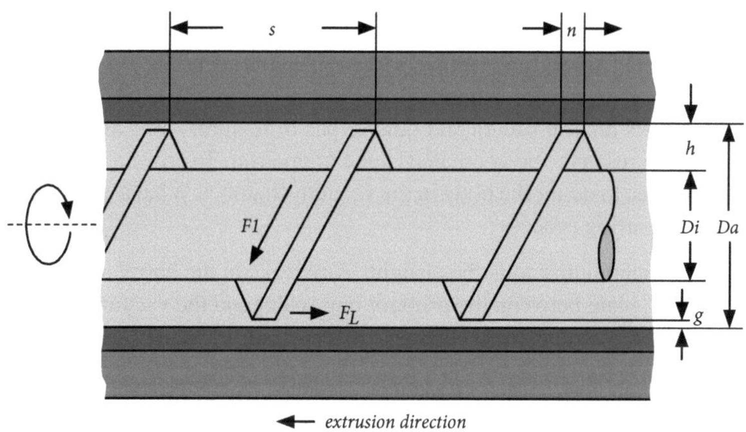

Fig. 2.8 Geometry of the auger

Augers can be characterized by the following main features:

- Geometry: length of the auger L, diameter D_a, diameter of the hub D_i, auger pitch s, ratio L/D_a, ratio D_a/D_i
- Further change of the pitch, the outside and/or inside diameter of the auger along the length of the auger
- Quality: type of material, wear-resistant coating, surface roughness, tempered or non-tempered
- Top auger: single wing or multiple wings; other systems such as "wiper".

2.5.2 Barrel

The barrels can be characterized in a similar way; this is where the so-called pressure build-up and compaction of the body are affected—see also Chap. 4—from the infeed zone into the enclosed barrel.

- Geometry: length (*L*), inside diameter (*D*), gap between barrel and auger (g)
- Quality: with or without replaceable liners, material quality of the liners, type of grooves in liners (linear, spiral type, sawtooth pattern, etc.)
- Tempering: with or without tempering of the individual barrel modules.

In the book "Extrusion in Ceramics" (2007) under the chapter "Types of Extrusion Units" written by Willi Bender, you will find a very detailed description of the different extruder variations, respectively, vacuum units.

2.6 De-airing

If the aerated ceramic body is compressed, it is not just the body which is compacted; the air is compacted too and an explosion-like expansion occurs as soon as the ceramic column is exposed to the atmosphere after leaving the die. The consequence is that the column shows dragon's teeth and other types of textures—see also Chap. 9—in addition to the pores in the body created by the trapped air. To prevent these damaging effects, it is a must to de-air the body in the vacuum chamber. What are the key factors for a perfect de-airing process?

- The entire vacuum unit must be airtight! Any leaks in the barrel, at the bearings, at the sealing plate between the primary pug sealer and the vacuum chamber will not only cause vacuum loss, but will also result in unwanted drying—see also Chap. 13.
- The vacuum pump must be dimensioned accordingly to suit the throughput rate and the product; i.e. the suction capacity (m^3/h) and the so-called ultimate pressure (millibar/Pa) must match these requirements.
- It is important that chippings or pellets which enter the vacuum chamber are of a small cross section since the vacuum can penetrate only a few millimetres. A longer retention time in the de-aired chamber may help, however, be less effective because of the smaller cross section and ultimate pressure.
- The vacuum level depends on the specific atmospheric pressure at the site and on the prevailing temperature of the ceramic body. If this leads to water evaporation in the vacuum chamber, it will cause the material to dry out with all the associated undesirable consequences; for more information, see Sect. 11.5.

The required vacuum readings are usually between 20 and 100 mbar, but a too high level of vacuum may also be unwanted, in which case external air can be added with the aid of a so-called poppet valve to reach a reduced vacuum level.

2.7 Extruded Ceramic Products

David W. Richerson's very well-illustrated and easy-to-read book "The Magic of Ceramics" (2012) offers a good overview of the kind of ceramic products we find in everyday life. The chapters dealing with technical and advanced ceramic products are of particular interest. Although not all of these products are extruded, you will have learnt here in Chap. 2 that the extrusion process is one of the most important shaping methods of ceramic materials.

The range of extruded ceramic products covers bricks of extremely diverse geometrical shapes, moreover high-tech honeycombs featuring wall thicknesses of less than 0.2 mm, flue pipes as well as semi-closed dome-shaped pipes, SOFC fuel cells, liquid filters and rustic style floor tiles, billets which may weigh up to four tonnes for the production of insulators through to tiny trilobes weighing a few grams to act as catalysts in the petrochemical industry.

It would go beyond the scope of this booklet to describe in detail the many varied ceramic products made by the extrusion system; I would therefore like to refer to the relevant publications and also to Sect. 2.1 where an attempt has been made to highlight the significance of extrusion as a shaping system in ceramics: it is certainly one of the most important and commonly employed systems.

2.8 Conclusions

- Collect all available information relating to the specific characteristics of extrusion, including the pros and cons in comparison with other shaping systems in ceramics.
- Determine how many items are to be produced per unit of time and how complicated is the geometry. Bear in mind that extrusion is primarily suitable for mass production. But there are exceptions to the rule.
- Also give some initial thought to the process stages upstream and downstream of the extrusion, e.g. is the available ceramic material already of a plastic consistency, or must it first be made "extrudable"?

Bibliography

1. R. Bartusch, Stand und Trends der Extrusionstechnik in der Technischen Keramik. Keramischer Zeitschrift **62**, 340–344 (2010)
2. W. Bender, Types of extrusion units, in *Extrusion in Ceramics*, ed. by F. Händle (Springer, Berlin, 2007)
3. W. Benderx, From Brick machine to high-tech extruder - 120 years of Händle extrusion technology, in *ZI-Annual 2010* (1890–2010), pp. 2–48

4. F. Laenger, Evacuation in ceramic extrusion—dependencies and local situations, in *Extrusion in Ceramics*, ed. by F. Händle (Springer, Berlin, 2007)
5. C.O. Pels Leusden, The basics of shaping, in *Brick and Tile Making*, ed. by W. Bender, F. Händle (Bauverlag GmbH, Wiesbaden und Berlin, 1982)

Chapter 3
Ceramic Materials—The Malice of the Object

The map is not the territory.
—Alfred Korzybski

Abstract Ceramics are classified into two schemes according to material and applications. The main properties of ceramic bodies for extrusion like yield point, viscosity are explained.

Keywords Ceramic materials · Applications of ceramics · Bingham fluid · Properties of ceramic bodies

3.1 Definition of Ceramics

> We define ceramics as the art and science of making and using solid articles which have as their essential component, and are composed in large part of, inorganic non-metallic materials.

This definition of ceramic stands at the beginning of the famous classic "Introduction to Ceramics" (2011) by Kingery et al. More details can be gathered from the basic literature mentioned in the references in Chap. 1. The detailed characterization of the various ceramic materials cannot be covered within the scope of this booklet, particularly in view of the fact that there is an excellent and vast literature on this issue; see also Wikipedia under the heading "Ceramic" including the associated links. The definition given above does not indicate in which physical state—solid or liquid—and form—compact or bulk—ceramics exist. Therefore, we refer to a different definition of ceramics, following a publication of Kriegesmann and Kratz, according to which a ceramic is a compact, synthetic and solid material always being polycrystalline and possibly containing some glassy phase. Ceramics receive their state through a high-temperature exposure. In order to get orientation in the confusing ceramic jungle, let us use two different but supplementary schemes.

F. Händle, *The Art of Ceramic Extrusion*, https://doi.org/10.1007/978-3-031-63471-0_3

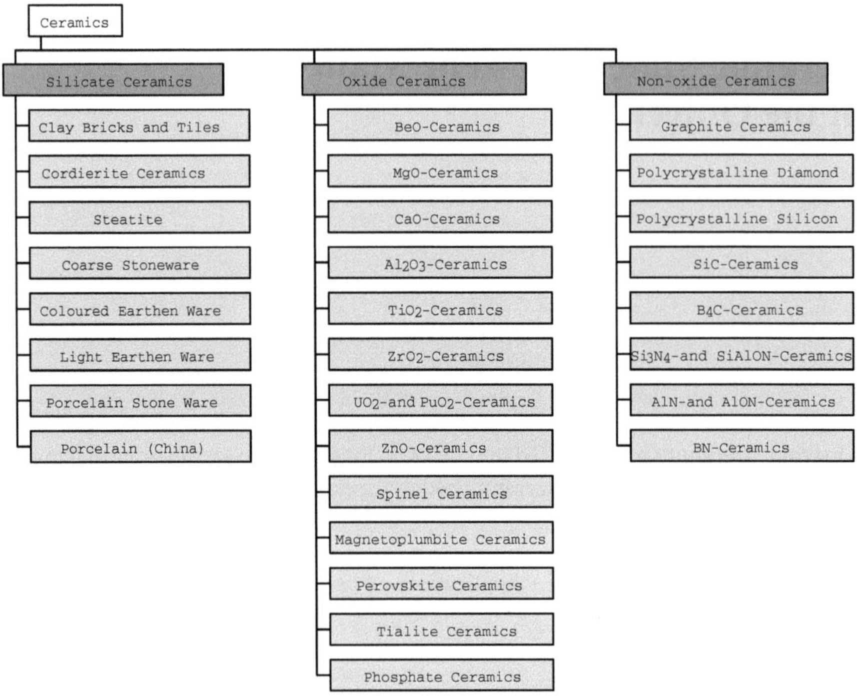

Fig. 3.1 Classification of ceramic materials

3.2 Classification of Ceramic Materials

In the first classification scheme—see Fig. 3.1—the specific ceramic materials are classified into three groups: silicate ceramics, oxide ceramics and non-oxide ceramics. The further subdivision of the authors is abandoned here.

3.3 Classification of Ceramics According to Applications

In the second classification scheme—see Fig. 3.2—ceramics are classified according to aspects of applications in compressed form.

Several attempts have been made to combine the two classification schemes such as the worth-reading contribution of Reh—see bibliography. A combined classification would, however, be confusing on the one hand in the light of the huge variety of ceramic materials and on the other since many ceramic materials are used for different applications. This is also contradicted by the fact that ongoing changes occur through the permanent technical progress in materials development and through opening up

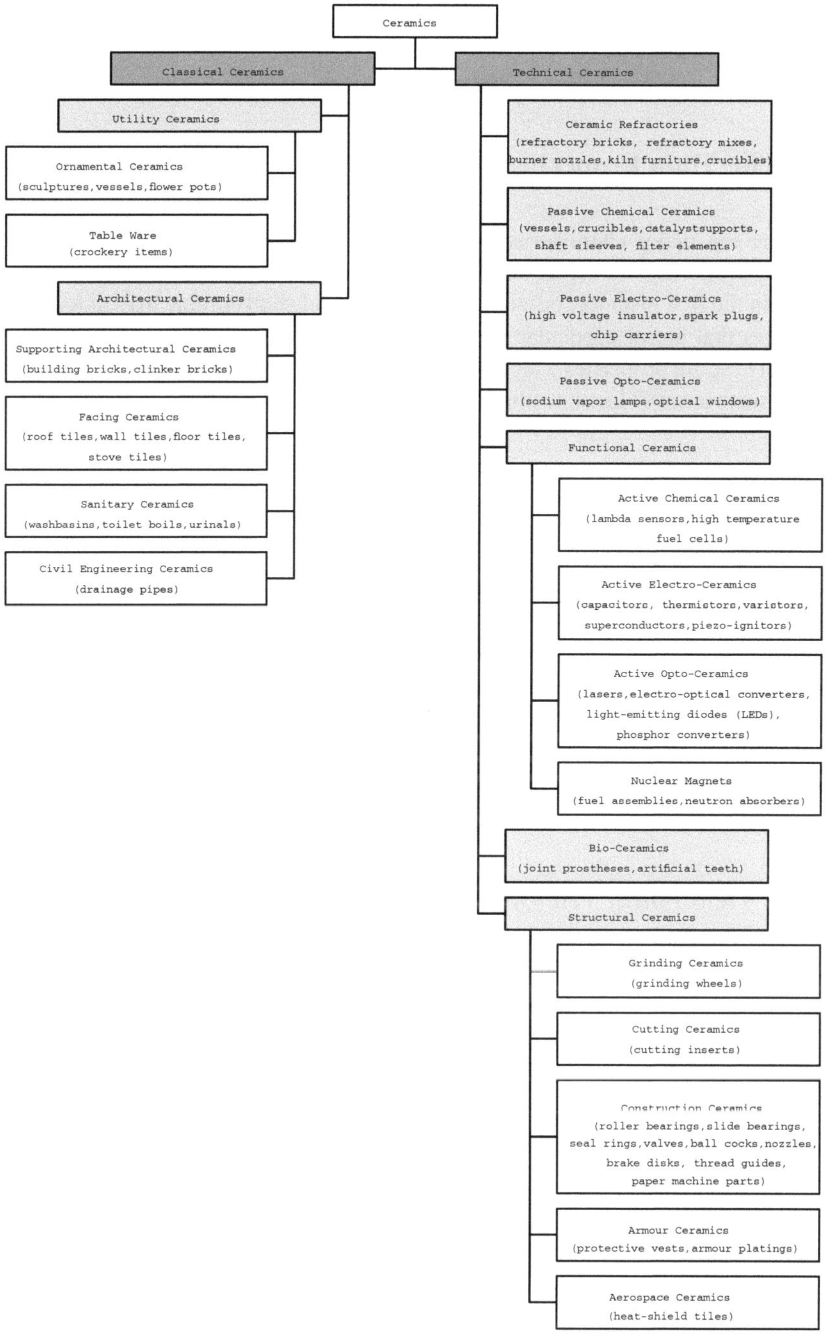

Fig. 3.2 Classification of ceramics according to applications

ever new applications for ceramic materials, which makes the validity of such a complex scheme doubtful.

3.4 The Main Properties of Ceramic Bodies for Extrusion

What are the major characteristics of ceramic materials used for extrusion? The "primary properties" of the mineralogical structure governed by the "formula" must, at least in the first instance, be assumed as given. So I shall confine myself to the key properties that are of particular relevance to extrusion from a process-engineering aspect. These are properties which we can influence by means of process-engineering "adjustment screws" like kneading, dosing, ageing, etc.

3.4.1 Plasticity

Owing to the fact that ceramic materials can only be extruded in a plastic state, it makes sense to consider plasticity as the outstanding feature of ceramic bodies to be extruded. Dry non-plastic powders cannot be extruded; they can only be shaped by dry or isostatic pressing. Ceramic slurries are also unsuitable for extrusion; different casting methods are employed instead for shaping purposes—see also Chap. 2. This means that plasticity is a "*conditio sine qua non*" for bodies to be extruded. For more information, please refer to Chap. 6.

3.4.2 Yield Point, Viscosity and What Mr. Bingham Has to Say

Perhaps you remember from your school and student days what is understood to be a Newtonian fluid. This is a liquid such as water with linear viscous flow properties. In other words, a liquid or compound for which the course of the shear stress is proportional to the shear rate. These cases are also referred to as shear-rate-independent viscosity. In comparison, non-Newtonian fluids/compounds change their viscosity under stress. Broadly speaking, they either become thin or viscous the more they are exerted to stress; i.e. their viscosity is shear rate-dependent. A distinction is made between two basic types: dilatant behaviour and structural viscous behaviour. In the case of dilatant behaviour, the shear stress increases disproportionately with increasing shear rate, referred to as "shear thickening". With structural viscous behaviour, it is vice versa and is called "shear thinning". A further aspect to be considered is that many non-Newtonian compounds start to flow only when exposed to a certain stress as is the case with plasticine, soft cheese or clay used for pottery.

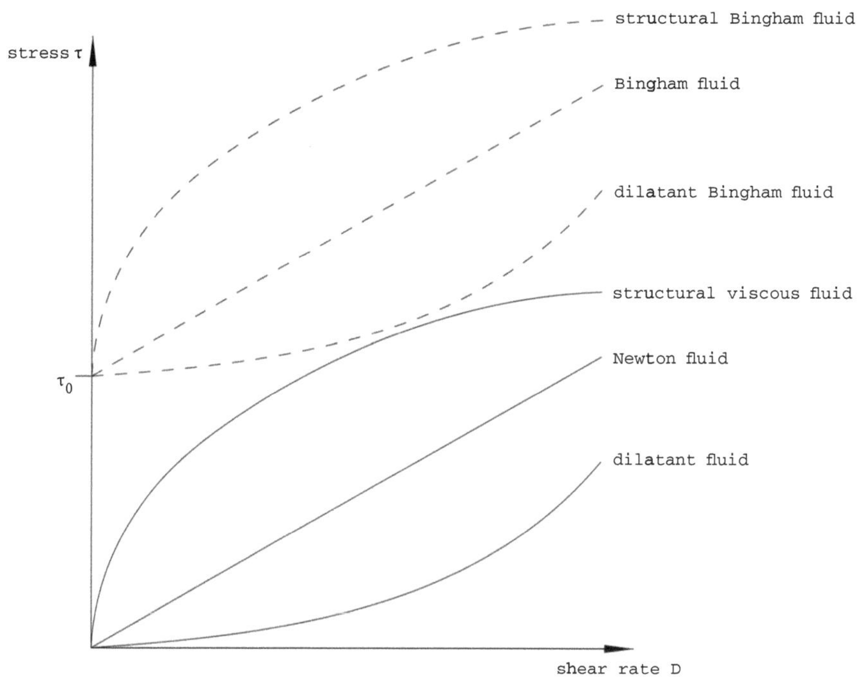

Fig. 3.3 Behaviour of ceramic bodies—shear stress versus shear rate

The stress limit is the "yield point" τ_0. With a linear course, such fluid or compound is called "Bingham fluid/compound", named after the well-known American rheologist Eugene C. Bingham, who in 1922 published his classic book "Fluidity and Plasticity" (Fig. 3.3).

Have you ever come across ceramic materials that change their viscosity in the course of time under constant stress? These are thixotropic or rheopex compounds. If you are interested, you could get more information on this subject from the relevant literature; please look up references in Chap. 1.

What you suspected a whilst ago must now be obvious: there are quite malicious ceramic materials that due to their complexity are difficult to handle. The German philosopher F. T. Vischer named this behaviour the "malice of the object". A truly amazing prediction in the context of today's "Internet of things". However, we also find "good-natured" ceramic bodies which forgive our faults; moreover, we have learnt by now how to outsmart less good-natured ceramic materials.

Fig. 3.4 Testing of bending strength

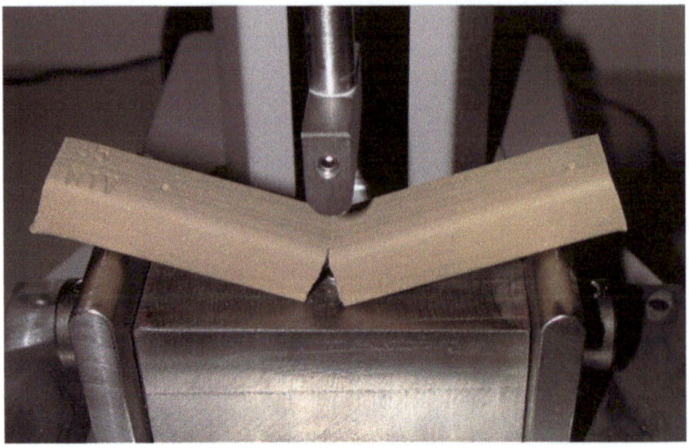

3.4.3 Homogeneity

The most important information on the subject of homogeneity can be found in the chapter on mixing and kneading; see Chap. 7. Accurate dosing of all components is the first condition to achieve a homogeneous body. Secondly, it is a must that these components meet the specified characteristics with respect to grain size distribution, moisture content, etc. In practise, these requirements are rarely met even in situations where predominately synthetic components are used. It is clearly not necessary to be an expert to come to the conclusion that the extrusion of inhomogeneous materials results in inhomogeneous extrudates. Unfortunately, also in practise, this understanding is not always prevalent, and as a result, the solution to problems is searched for in the wrong place.

How can we determine whether a body is homogeneous or inhomogeneous? The most proven method is still by measuring the bending strength of dried samples produced from the compound to be tested—see Fig. 3.4. This is a virtually indirect yet proven system as long as enough readings are taken in order to obtain representative data; see also Chap. 15. The addition of luminescent powders during mixing to examine the statistical distribution with the aid of ultraviolet light has also been adopted in practise to check the mixing quality. A less known method, but yet very meaningful in my opinion, is the use of a capillary rheometer for measuring the pressure fluctuations, from which the degree of homogeneity of the mixed or kneaded material can be deduced.

3.4.4 Grain Size Distribution

When we talk of homogeneous materials, it is assumed that we refer to materials with a constant homogeneity over time, i.e. constant in terms of moisture content, mineral structure, inert and organic components, grain size distribution, etc. It is a requirement that an ideal or desired grain size distribution exists for each ceramic compound, and this should be maintained within a more or less narrow corridor. Failing this, because the proportion of coarse and fine grains varies disproportionately, a constant homogeneity cannot be assumed even though the compound is mixed and kneaded properly. Moreover, it must be borne in mind that the grain size distribution is influenced by process-engineering measures in contrast to the mineral structure. In course of the material preparation, e.g. during mixing and kneading, also whilst straining and extruding, the particles are subjected to crushing and grinding, dispersion, agglomeration or de-agglomeration. These may be target-oriented processes such as an additional shearing action in order to increase the percentage of a certain grain size range, or unwanted processes such as undesired agglomeration of particles.

3.4.5 Wall Friction

In extrusion, we distinguish between the "internal friction" of the material and the "outer friction", also called "wall friction". Whilst the viscosity and yield point of the body can be determined relatively easily, it is quite elaborate to empirically establish the wall friction, all the more so as this varies with the pressure rate. The fact that the wall friction depends on the roughness and condition of the "wall" which could be the auger, the die, etc. on one hand, and on the other, the consistency or viscosity of the material, which can by no means be assumed as a constant, does not make things easier. Three major variables are needed to simulate the flow of ceramic materials: the viscosity of the body, the yield point and the wall friction. Whereas computational fluid dynamics (CDF) simulations of complex profiles failed in the past due to the limitations in the performance capacity of the models or the calculating speed of standard computers, the present problems lie in the empirical determination of the wall friction within a reasonable time and cost frame. It is hoped that this problem can be solved in the foreseeable future with the aid of new approaches such as that of Laenger et al.

3.5 Conclusions

- Obtain sufficient information regarding the properties and characteristics of the ceramic material in terms of process-engineering options.
- Check if the primary properties of the ceramic body fit to the extrusion of the required product.
- In case of problems, check if you can optimize suboptimal properties by modifying the available "adjustment screws".

Bibliography

1. T. Haase, *Keramik* (VEB Deutscher Verlag für Grundstoffindustrie, Leipzig, 1967)
2. J.G. Heinrich, C.M. Gomes, in Introduction to ceramic processing. www.ceramics-processing.com (2017). Accessed 23 April 2018
3. W.D. Kingery et al., *Introduction to Ceramics* (Wiley, New York, 2011)
4. J. Kriegesmann, N. Kratz, Definition, Systematik und Geschichte der Keramik: Von der Tonkeramik bis zur Hochleistungskeramik. Keram. Z. **67**, 152–157 (2015)
5. J. Kriegesmann, N. Kratz, Definition, Systematik und Geschichte der Keramik: Einteilung der Keramik nach werkstoffspezifischen Gesichtspunkten. Keram. Z. **67**, 227–230 (2015)
6. J. Kriegesmann, N. Kratz, Definition, Systematik und Geschichte der Keramik: Einteilung der Keramik nach anwendungsorientierten Gesichtspunkten. Keram. Z. **67**, 276–287 (2015)
7. F. Laenger, et al., in Wall slip of ceramic extrusion bodies, Part 2. cfi 93, E1–E5 (2016)
8. F. Laenger, Rheology of ceramic Bodies, in *Extrusion in Ceramics*, ed. by F. Händle (Springer, Berlin, 2007), pp. 141–159
9. H. Reh, Current classification of ceramic materials, in *Extrusion in Ceramics*, ed. by F. Händle (Springer, Berlin, 2007), pp. 35–58
10. R. Telle (ed.), *Salmang/Scholze Keramik* (Springer, Berlin, 2007)

Chapter 4
A Little Bit of Theory

> *There is nothing more practical than a good theory.*
> *—Immanuel Kant*

Abstract Two theoretical approaches within the framework of "extrusion theory" for ceramic materials are explained. There is also a table based on the formula for calculating the throughput capacity, which lists the throughput rates for those extruders commonly used in the ceramics industry. Definitions of the key basic terms necessary for a better understanding of the theoretical approaches are summarized.

Keywords Pressure generation · Backflow · Benbow–Bridgwater equation · Extrusion output · Shaping degree

4.1 What This Chapter Is About

A bit of theory does help, but no worries, everything is explained quite clearly just spiced up with a minimum amount of mathematics. I shall limit myself in all brevity to describe two approaches relating to "extrusion theory" to be followed by a simple scheme with the aid of which you can with sufficient accuracy calculate the throughput capacity of your extruder. First of all, I shall try to explain the pressure build-up in an extruder, whereby I shall largely fall back on the information provided by Pels Leusden and Lutz. Then is an introduction to the Benbow–Bridgwater approach to the calculation of pressure rates. Every extruder can also be understood as a kind of continuous or intermittent pump, i.e. a device that feeds/extrudes fluids or pastes against a resistance, whereby in our case we are concerned with more or less incompressible media. For clearer understanding, we tend to speak of the extruder as a "pressure generator" that pushes against a resistance which we call the "pressure consumer". The amount of pressure consumed depends to a great extent on the viscosity of the material, the geometry of the die and the speed at which the material is fed through the die—so much for a start.

Try to imagine we are extruding a soft ceramic compound as used by the potters or, alternatively, a stiff hard metal material. Under otherwise identical conditions,

F. Händle, *The Art of Ceramic Extrusion*, https://doi.org/10.1007/978-3-031-63471-0_4

the two diverse materials would create totally different conditions for the pressure generator and the pressure consumer. I would love to get lost in the details, but this is precisely what I want to avoid.

4.2 Basic Terminology

Before we go deeper, there are five basic terms to be considered:

- **Pressure generation of the auger**

By this, we understand the capability of the auger to generate pressure as a function of its geometry, assuming the existence of plastic body. The most important parameters are the relationship between length and outside diameter L/D, the relationship between hub diameter and outside diameter of the auger D_i/D_o as well as the auger pitch s. For ceramic extruders, the L/D ratio is between 3 and 15 depending on the application; the D_i/D_o ratio is between 0.3 and 0.8.

- **Backlog point RSP**

By this, we understand that point within the barrel where the material, either supplied in bulk or in the form of pellets, is compacted and has almost reached the maximum green density.

- **Coefficient of pressure increase of the body**

By this, we understand the pressure which can be generated for each centimetre of auger length with a given auger using a specific ceramic body. Let us call this factor PG (expressed in bar/mm). Although this factor is extremely helpful in characterizing ceramic bodies both in the laboratory and also in practise, I have so far never come across this term in the pertinent literature. Let us imagine you generate a pressure of 60 bar on a laboratory extruder using an auger of 300 mm length with a stiff ceramic body and a given die; i.e. $PG = 0.2$ bar/mm. Now you repeat the same test with a slippery soft body and you can only reach a pressure of say 30 bar, i.e. $PG = 0.1$ bar/mm. Just consider the consequences this may have in terms of practical operation. In Fig. 4.1, this factor is represented by the angle alpha.

- **Degree of re-shaping**

By this, we understand the natural logarithm of the outlet cross section of the augers or that of the piston extruder in relation to the cross section of the die. As a simple example, let us take an extruder with a barrel diameter of say 150 mm producing a compact rod of say 40 mm in diameter, i.e. $\ln(17{,}662/1256) \equiv 2.64$. Empirical tests with different materials have shown that sufficiently accurate forecasts can easily be achieved by adopting this approach using dies of simple geometry.

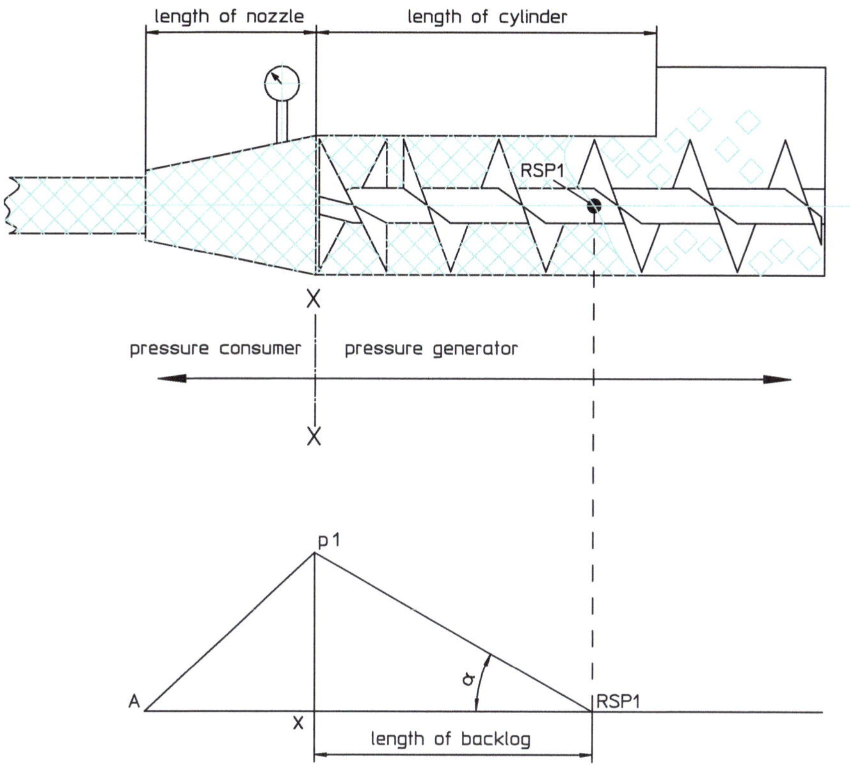

Fig. 4.1 Development of pressure in extruder

- **Memory of the ceramic body**

Not only do we humans have a memory, but ceramic bodies have one too. By this, we understand the plastic resilience of the ceramic body, which in particular can be observed during the drying process "whereby the particles try to find their energetically favourable location in direction of their initial location prior to the re-shaping"—at least this is the definition in a relevant encyclopaedia. At this point, you may be starting to recall a quotation by Richard Feynman: "If you thought that science was certain—well that is just an error on your part."—or perhaps thinking esoteric pseudoscience. But both positions would be in this case plainly wrong! Everyone involved in working with ceramic bodies is aware of the memory phenomenon, and it has also been the subject of scientific study. More on this subject can be found in the relevant literature.

4.3 Scenarios of Pressure Build-up in the Extruder

In order to explain the pressure build-up in an ideal-typical extruder, we take an extruder with a certain auger geometry, a ceramic body with defined rheological properties such as viscosity, wall friction, a given die and a predetermined throughput rate. The required pressure rate for extrusion (which we call *P1*) can be derived from this. The pressure build-up in the extruder can be represented clearly in graph form; see Fig. 4.1. The *x*-axis shows the auger from the infeed point to the die. The *y*-axis shows the generated pressure rate of the ceramic body over the length of the auger. De-aired loose material is fed into the auger from the infeed area. This loose material gets gradually compressed into a compact mass in the pressure zone. The so-called backlog point *RSP1* defines this state of compaction. The highest pressure is reached just ahead of the so-called end or top auger which, depending on the geometry, splits the column from the single auger into several columns, usually two or three. The pressure starts decreasing already in the area of the end auger and continues to do so in the pressure head and then drops down to zero after exiting the die. However, for the sake of simplicity, the required pressure rate P1 will be assumed as a maximum at the end auger in the further course. The angle alpha defines the coefficient of pressure increase of the ceramic body. In the following, I would like to discuss three relevant scenarios.

4.3.1 Scenario 1: The Notorious "Backflow"

What would happen if the pressure generator cannot produce the higher pressure *P2* required by the pressure consumer because the geometry of the die has been altered whilst the ceramic body remains identical, be it that the degree of re-shaping has increased and/or the wall friction of the die has gone up? This situation can be seen in Fig. 4.2. With the same coefficient of pressure increase of the material characterized by the angle alpha, the pressure build-up ability of the auger is not high enough. The material flows along the auger flights and through the gap between the auger and barrel back into the infeed zone. This means that the backlog point *RSP2* is moved from the enclosed pressure zone back into the open infeed zone where no pressure can be generated. This phenomenon is well known by the term "backflow". The next scenario will show that there could be other reasons for backflow.

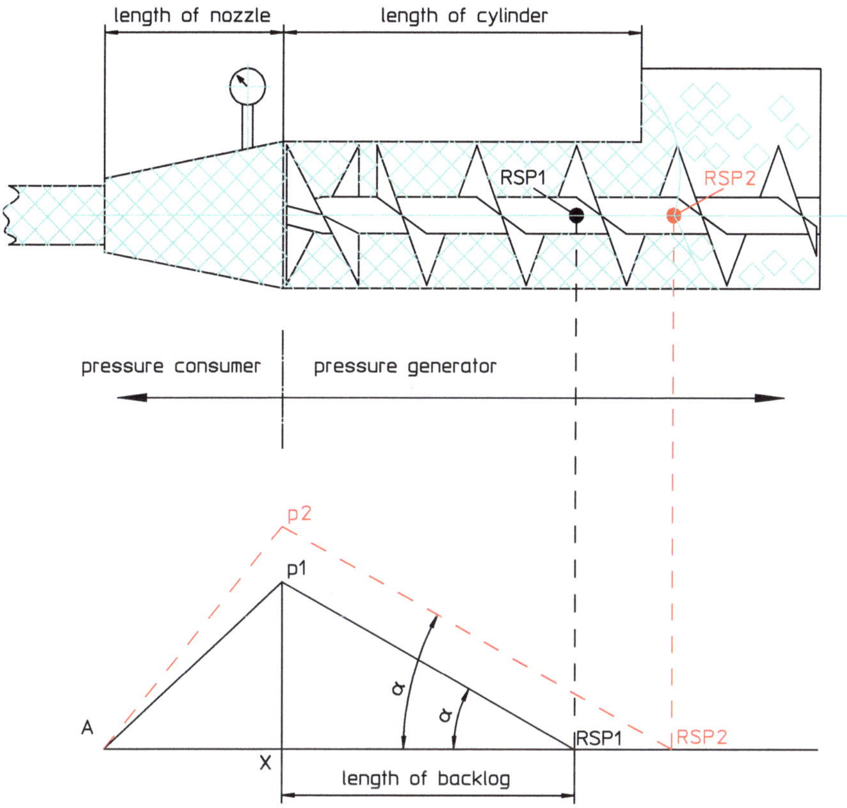

Fig. 4.2 Development of pressure in extruder—shifting of backlog I

4.3.2 Scenarios 2 and 3—"Backflow" Again, But for Other Reasons

What happens if the required pressure drops to *P3* because the body became softer in line with the coefficient of pressure increase beta? This could also produce a backflow situation as shown in Fig. 4.3. Due to the smaller angle beta, the backlog point falls back to the open infeed zone now corresponding to backlog point *RSP3*. Only if the body gets even softer corresponding to a pressure *P4* or a coefficient of pressure increase shown by angle gamma, is the auger length adequate. The backlog point *RSP4* is now in the pressure barrel.

Other scenarios such as higher throughput rate or combined scenarios with variations in the properties of the ceramic body or degree of re-shaping can be graphically deduced from these graphs. As regards the backflow phenomenon, it should be added that this not only entails quantitative consequences such as reduced capacity, but also quality problems due to the fact that the ceramic body is sheared several times during a backflow situation, which may influence the viscosity of the body. Moreover, in

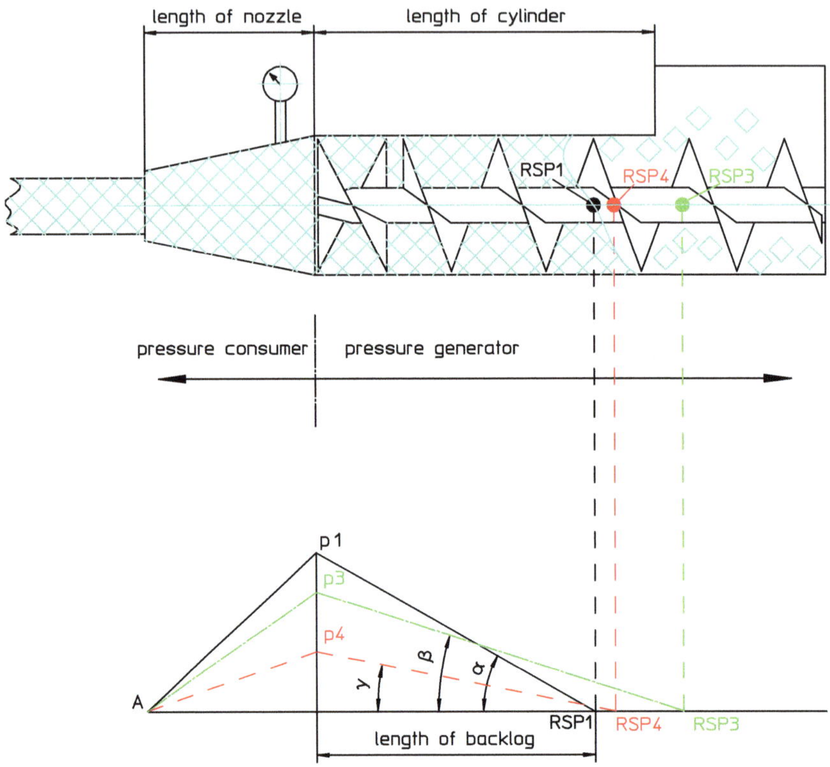

Fig. 4.3 Development of pressure in extruder—shifting of backlog II

certain cases, heightened wear could be encountered and increased contamination of the body. For these reasons, it is imperative to avoid backflow.

4.4 The Benbow–Bridgwater Equation

An article by Benbow, Bridgwater et al. under the title of "Prediction of Paste Extrusion Pressure" was published in 1989 in the Ceramic Bulletin. With the aid of this model, later known by the term "Benbow–Bridgwater Equation" (BBE), it is possible to estimate the expected pressure rates for plastic extrudable materials with sufficient accuracy. It is however a condition that the material-specific properties on which the formula is based can be identified or assessed empirically. This is the formula:

$$p = \left(\sigma_0 + \alpha \cdot v^m\right) \cdot \ln\left(\frac{A_0}{A}\right) + \left(\tau_0 + \beta \cdot v^n\right) \cdot \left(\frac{4L}{D}\right)$$

In the meantime, this model has in numerous contributions, in particular, those provided by the Department of Chemical Engineering and Biotechnology, Cambridge, been substantially refined, extended and adopted for the most diverse bodies. In his well-known textbook "Ceramic Processing", J. Reed also makes reference to it.

For readers who are not well versed in mathematics, here is a brief explanation of what this is all about and why we need to deal with this. The BBE formula assumes that the total pressure p can be determined by adding the two terms $ps = (\sigma_0 + \alpha \cdot v^m) \cdot \ln(A_0/A)$ and $pr = (\tau_0 + \beta \cdot v^n) \cdot (4L/D)$, whereas ps signifies the shaping/deformation process whilst pr signifies the friction process. Besides the body-specific parameters such as τ_0, σ_0, α *and* β, the relevant variables for ps are the extrusion speed v^m and the degree of shaping/deformation described by the term $\ln(A_0/A)$; pr is determined also by the extrusion speed v^n and the term $(4L/D)$ which describes the geometry of the pressure head/die combination. Again in simple words: the generation of the pressure is determined by the characteristics of the body, the geometry of the pressure head/die combination, the shaping degree the extrusion speed and a factor for the friction in the pressure head/die combination.

With complex profiles such as honeycombs, the term *(4L/D)* must be extended by a factor for the friction in the pressure head and die. Let me explain this with a simple example. In Fig. 4.4, you see two shapes with the same free cross section, or in other words, with the same shaping factor. But obviously, the friction area of shape B is much higher than the friction area of shape *A*. Assuming both shapes—or let's call it dies—have a length of 10 mm, the friction area of shape A would be 1865 mm^2 whilst the friction area of shape B has a much higher friction area of 39.040 mm^2.

As an alternative, it would be possible to empirically establish in the laboratory a so-called die coefficient for a specific die and a specific body as suggested by Pels Leusden. If you are interested in the BBE approach or the suggestions of Pels Leusden, please consult the relevant literature listed under references in this chapter.

4.5 How to Calculate the Output of an Extruder

Of course, you must be interested in knowing the output which can be achieved with a certain type of extruder. Here is the formula for the throughput expressed in litres per minute:

$$V = \eta \cdot n \cdot \pi \cdot \frac{\left(D_a^2 - D_i^2\right)}{4} \cdot (s - h) \cdot \frac{1}{10^6}$$

V Throughput rate [l/min]
η Extrusion efficiency [%]
n Revolutions of the auger [rpm]
D_a Auger outside diameter [mm]
D_i Auger inside diameter [mm]

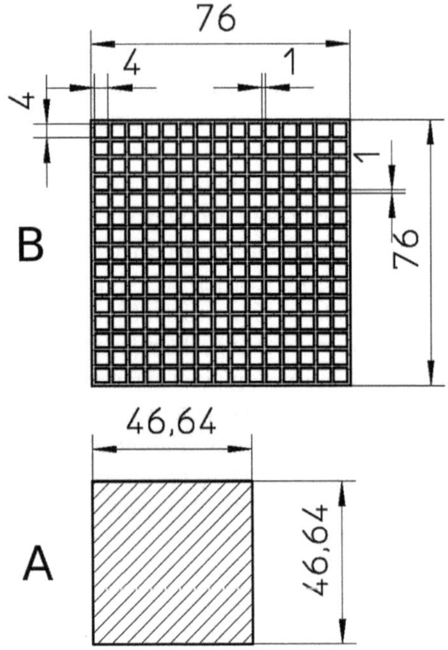

Fig. 4.4 Friction area of two shapes with identical free cross sections

s Pitch [mm]
h Thickness of auger flights [mm].

Whereas the geometrical data of the auger can be determined without any problem, the extrusion efficiency η depends on many boundary conditions such as state of the auger and barrel, properties of the ceramic body such as its viscosity, wall friction, degree of shaping, the pressure rate, extrusion speed and the adequate feeding of the auger. It must also be pointed out that a higher speed will not give a linear increase in throughput. On the contrary, if a maximum reading is exceeded at too high speeds, it can drop to zero. Goodson has provided some important contributions on this subject.

We have devised a table to facilitate the assessment of the throughput capacity, which can be applied for this purpose as long as what has been said above is taken into account and the boundary conditions assumed by us are taken into consideration. We have assumed auger diameters ranging from 50 to 700 mm, a ratio of hub diameter to auger diameter D_i/D_a of 0.5, an average pitch angle of 17° and an extrusion efficiency η of 40% corresponding to 0.4. The thickness of the auger flights was adjusted to suit the auger diameters. The discharge capacities expressed in litres per minute (in rounded figures) are listed for the different auger dimensions as a function of the speeds (Table 4.1).

Table 4.1 Throughput (L/min) for different diameters and auger revolutions

Diameter (mm)	Auger revolution (rpm)					
	5	10	15	20	25	30
50	0.1	0.2	0.3	0.4	0.4	0.5
100	0.7	1.4	2.1	2.8	3.5	4.2
150	2.5	4.9	7.4	9.9	12.3	14.8
200	5.9	11.9	17.8	23.8	29.7	35.6
250	11.9	23.9	35.8	47.7	59.7	71.6
300	20.8	41.6	62.4	83.2	104.0	124.8
350	33.5	67.0	100.5	134.0	167.5	201.0
400	49.6	99.2	148.8	198.4	248.0	297.0
450	71.4	142.7	214.1	285.5	356.8	428.0
500	98.7	197.4	296.2	394.9	493.6	592.0
550	132.3	264.6	396.9	529.1	661.4	793.0
600	170.6	341.2	511.8	682.3	852.9	1023.0
650	218.1	436.3	654.4	872.5	1090.6	1308.0
700	273.8	547.5	821.3	1095.1	1368.9	1642.06

4.6 Conclusions

- Ensure that you do understand the few basic terms mentioned in this article.
- Address the two theoretical approaches. They are very useful for problem-solving.
- Calculate the feed efficiency η for your extruder for different dies and ceramic bodies.

Bibliography

1. J. Benbow, J. Bridgwater et al., Prediction of paste extrusion pressure. Ceram. Bull. **69**, 1821–1824 (1989)
2. F.J. Goodson, Experiments in extrusion. J. Trans. Br. Ceram. Soc. **58**, 158–187 (1959)
3. F. Händle, F. Laenger, J. Laenger, Determining the forming pressures in the extrusion of ceramic bodies with the help of the Benbow-Bridgwater equation using the capillar check, in *cfi* (2015), pp. E1–E5
4. J.G. Heinrich, C.M. Gomes, in Introduction to ceramic processing. www.ceramics-processing.com (2017). Accessed 23 April 2018
5. D. Lutz, Scenarios of extrusion, in *Extrusion in Ceramics*, ed. by F. Händle (Springer, Berlin, 2007), pp. 173–186
6. F. Laenger, et al., Wall slip of ceramic extrusion bodies—part 2, in *cfi 93* (2009), pp. E1–E5
7. C.O. Pels Leusden, Die Formgebung auf Strangpressen, *Ziegeltechnisches Jahrbuch* (Bauverlag BV GmbH, Gütersloh, 1975), pp. 237–379
8. J.S. Reed, *Principles of Ceramic Processing* (John, New York, 1995)

Chapter 5
How to Determine an Extruder and Which Data Do We Need?

We think, each of us, that we're much more rational than we are. And we think that we make our decisions because we have good reasons to make them. Even when it's the other way around. We believe in the reasons because we've already made the decision.
—Daniel Kahneman

Abstract To determine an extrusions plant, it is advisable to follow a structured incremental decision procedure and to have the basic data available regarding the required product, the foreseen output, the main characteristics of the body, etc. A well-proven decision process in six steps is presented as well as a basic questionnaire.

Keywords Decision procedure · Database and questionnaire · Kneading curve

5.1 The Decision Procedure and the Database

We are asked quite often which particular extruders would come into question based on throughput rate A, for products of type B, using ceramic body C, and appropriately, it is those three parameters that are essential for the initial selection of a suitable extrusion plant; or more precisely, these data are necessary but not sufficient. What could a decision procedure look like that would be transparent for both sides—the manufacturer of such extruders and likewise the client—and which even at a later date be fully comprehended?

Our approach consists essentially of two elements:

- Firstly of a six-stage decision model and
- Secondly of a basic questionnaire covering the most important questions that need to be answered for a first assessment.

In our opinion, a consistent and rigorous handling and evaluation of these two elements are to be considered a "must" for a rational selection, whereby economic considerations, such as price or financing benefits, are omitted. This does not mean that economic aspects should be neglected—of course not—but they should not be

F. Händle, *The Art of Ceramic Extrusion*, https://doi.org/10.1007/978-3-031-63471-0_5

taken into consideration for the "first round" which is intended for assessment of the technical criteria. Some of the questions listed in Sect. 5.3 may be difficult to answer especially for those who have not been sufficiently acquainted with the subject thus far, but unfortunately, we have to live with some lack of information, as is typical for any kind of decision-making, as long as one realizes that these deficits exist.

Open questions will be answered after each of the five steps along with new questions that might arise, whereby it may happen that initial ideas, either of a qualitative or quantitative kind, need to be revised. Incidentally, the incremental and systematical approach in the decision-making process is the best guarantee that it will not come to irreversible decisions because the process can be stopped after each individual step. This also guarantees that the original assumptions can be revised or modified at any time. Moreover, it follows from the very nature of the decision-making process that all persons involved, i.e. the client, a possible partner and consultant, as well as the extruder manufacturers, can communicate, advise and decide jointly as partners and equals, rather than against each other. You are quite right in thinking that this decision-making method is quite painstaking, but believe me, it is worth the effort (Fig. 5.1).

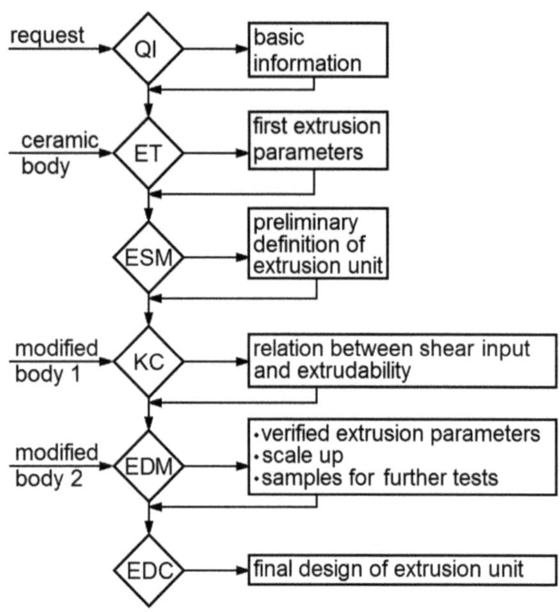

Fig. 5.1 Our decision model

5.2 The Decision Model

The model consists of six steps:

1. After receipt of your enquiry, we would send you a questionnaire—see Sect. 5.3—requesting you to complete the same as fully as possible and—if at all possible—to send us a sample of your material; we call it step QI.
2. By carrying out an extrudability test—ET—we determine whether it is basically possible to extrude your material and, in the affirmative, under which conditions. This test is performed with the aid of the "Capillar Check"—described more closely in Chap.6. During the course of this test, the detection of the essential extrusion parameters for your body with respect to the intended extrudate(s) is possible.
3. Assuming the ET produces positive results, in the third step, we are then able to define the basic features for the extrusion unit, such as torque of the gearboxes, bearings, auger geometries—or piston diameters—using the ExtruderSimulation Model—ESM.
4. The kneading curve KC—see Chap. 6—was developed by ECT in 2009. KC is a unique method to detect and optimize the relationship between the kneading/shearing/dispersion process and the extrudability/plasticity of the ceramic body. Universities, institutes and well-known companies all over the world are now using this method to optimize the body and/or the preparation process.
5. During the next step called the Extrusion Determination Method—EDM—we carry out various extrusion tests in distinguished laboratories on the representative material samples you have provided. The extrudates are normally generated with 80 mm auger extruders or a piston press. All relevant parameters like pressure rates, temperatures, torques, backlogs, linearity with respect to extrusion speed, shear stress, shaping factors, etc. are detected under real conditions. On the basis of these results, one is able to generate a proper "scale up" for the extrusion unit of the intended production plant.
6. The last step Extruder Design Criteria—EDC—is undertaken only once all the relevant and basic information—including the results of the tests and the final requirements of the customer—are available and the decision to go ahead has been made. On the basis of an extremely detailed checklist, all the relevant features of the extruder in question are double-checked. Think of the checklist, a pilot has to go through before he starts the airplane: the EDC procedure can be compared to that.

You might want to know how time-consuming the various steps are. For step 2, we need about one long day; steps 3 and 4 take about five days in total including the preparation of the report. For step 5, three to four days are required. Step 6 is an internal process and takes from two days upwards to several days if modifications of the extrusion unit are necessary.

5.3 About the Sense and Nonsense of Questionnaires

Have you also had enough of people who send you incredibly long questionnaires and tell you that completion will only take five minutes and how important it is to know your opinion? I am on your side, but I must also admit the truth that completing the questionnaire—Table 5.1—is crucial if you are seriously thinking about investing in an extrusion plant. The more accurately you can answer our questions, the more precise will be the response. This allows on the basis of the data provided to adequately implement the decision process without knowledge of the details of your body which is regarded as your intellectual property. However, only through the next steps—ET to EDM—are we able to verify whether our assumptions are correct. Should you have any laboratory test or other test results available, these should be sent to us along with the completed questionnaire and a clear statement regarding the necessity of a NDA which protects both partners.

Table 5.1 Example of a questionnaire as database to determine extrusion plants

General data	Your answer	Our remarks
Complete address of your company		
Name of person in charge		
Your phone number		
Your email address		
General questions		
Do you already produce similar ceramic products?		If yes, please send us a description.
What should be the exact scope of our offer?		Budget offer? Including preparation?
Confidential agreement required?		
Local conditions		
Max. temperature		
Min. temperature		
Max. relative humidity		
Estimated automation degree		
Height production plant building		
What kind of preparation is intended/available		

(continued)

Table 5.1 (continued)

General data	Your answer	Our remarks
Product data		
Type of products		
Dimensions (LxWxH) mm		All dimensions of all products, if necessary add drawings
% of perforation		
Weight of the green product (kg)		
Weight of the dry product (kg)		
Weight of the fired product (kg)		
Output Data		
Output/year (pieces)		Output data for the different products
Output/day (pieces)		
Working days/year		
Working hours/day		
Material data		If available add laboratory reports regarding mineralogy, grain size distribution, plasticity, etc.
Type of material		
Dry shrinkage		
Firing shrinkage		
Major additives		
Abrasive body		
Corrosive body		pH—value if available
Electrical data		
Voltage (V)		
Frequency (Hz)		

In Chap. 18 dealing with the "optimal extruder", you find a checklist designed to be used prior to your final decision to help you choose which one of your shortlisted extruders features those characteristics which are relevant to you.

Our decision model should be looked at critically. Can you follow the logic relating to the sequence of the individual steps? You will find that we have consciously incorporated redundant loops in "double-checking" with a view to safely "taking off" and "landing" the decision-making process.

5.4 Conclusions

- Please try to formulate your requirements with the utmost accuracy and do not be afraid to leave open any unclear details. For instance, if you do not know the working times, or if the ceramic body has not been optimized as yet. It is not critical at this stage if it is too early for you to answer certain questions as long as you are aware that there still exist "blind spots" and information deficits.
- Please also check whether there are any boundary conditions that are not covered by your answers in the questionnaire, but which could be relevant for the design and planning of the extrusion plant.
- Please try to set up your own decision model. Which criteria are the most important for you and how would you weigh the importance of these requirements individually, so as to finally arrive at an evaluation and decision which is the most objective possible?

Bibliography

1. D.W. Taylor, in *Decision Making and Problem Solving*, ed. by J.G March. Handbook of Organizations. (Rand Mc Nally, Chicago, 1965), pp. 48–86
2. D. Kahneman, *Thinking, Fast and Slow* (Penguin, London, 2012)

Chapter 6
Plasticity or the Great Unknown

Plasticity is like honesty: it is hard to define, but it is associated with certain properties.
—Scott Blair

Abstract The two definitions of plasticity in ceramics and how to measure deformation pressure and rupture strength. Which factors determine the plasticity of a ceramic body? About two relevant methods to detect the consistency and extrudability of ceramic bodies and the relation of these terms with plasticity.

Keywords Plasticity · Deformation pressure · Rupture strength · Kneading curve · Extrudability

6.1 What Does Plasticity Mean in Ceramics?

What do we understand by the term "plasticity" within ceramics? We say: "not plastic enough" or "too stiff" or "too soft"—but what exactly do we mean? It was the English scientist Scott Blair, writer of an important book on the subject of rheology, who gave us the above-mentioned sarcastic quote. In the following, I will endeavour first of all to provide more accurate definitions of the term plasticity as formulated by Brownell and Haase. I will then go on to explain how we can measure plasticity, define the most relevant factors determining plasticity, present two empirical methods to describe the development of plasticity and finally try to differentiate between the term plasticity and terms such as "consistency" or "extrudability".

6.2 The Two Plasticities

Whenever the term plasticity is employed it should be clear what is really meant. Brownell from Alfred University and Haase from the University of Freiberg tried independently to find precise definitions. In his book, "Structural Clay Products" (1970) Brownell stated the following:

F. Händle, *The Art of Ceramic Extrusion*, https://doi.org/10.1007/978-3-031-63471-0_6

> "Plasticity is the ability of a clay-water mass at its maximum consistency to be shaped and to hold its shape after the forming forces are removed."

In other words, this means that we are dealing with two different physical quantities. Plasticity I is understood to be the force we have to apply to tear a green plastic column; Haase calls it "rupture strength". Plasticity II means the resistance against penetration of a solid object into the ceramic body; Haase calls it "deformation pressure". In a diagram, Haase has shown both variables in relation to the water content of certain ceramic body.

The *y*-axis in Fig. 6.1 shows the rupture strength, whereas the upper *x*-axis indicates the deformation pressure. In reality, we often face a trade-off situation which means that we have to decide which of the two plasticity values is more relevant for our product and process.

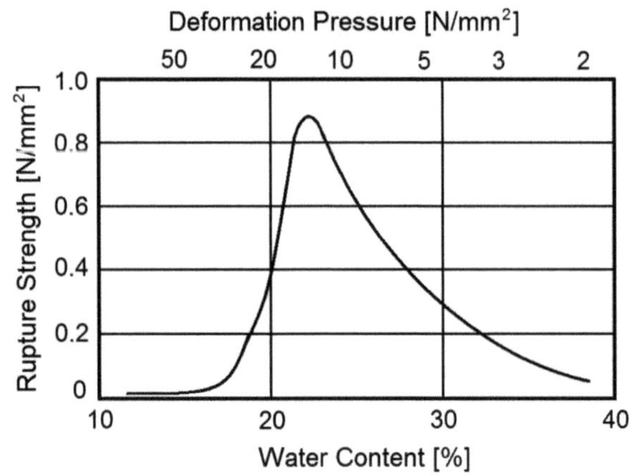

Fig. 6.1 Rupture strength versus deformation pressure

6.3 How Do We Measure the Two Plasticities?

Let me start with the good news: the deformation pressure can easily be measured by using variously designed so-called penetrometers (Fig. 6.2).

For practical purposes, this is quite adequate as long as certain rules are observed in measuring. To achieve more accurate, intersubjective readings, an instrument is recommended with which the penetration depth and a constant penetration speed can be preset. Both instruments measure a force, for instance Newton, from which the specific pressure, for instance in psi, bar or Pascal can be derived.

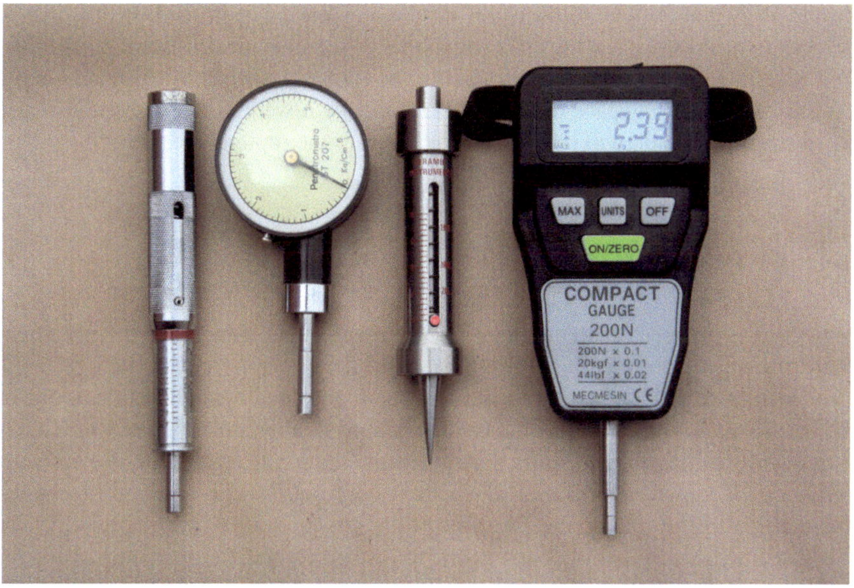

Fig. 6.2 Different types of penetrometers

Now the bad news: up to date no adequate measuring device is available to detect the rupture strength. In Fig. 6.3, I have tried to demonstrate what rupture strength

Fig. 6.3 A simple but not accurate way to measure the rupture strength

means. Why is it difficult to measure? Because in tearing apart the two-end, plastic deformation takes place at the ends which makes accurate measurement impossible. The best compromise to get an idea about the rupture strength would be to hang a ceramic rod freely in space until it breaks by its own weight. Let us hope that one day a suitable method will be available to measure the rupture strength.

6.4 Factors Determining Plasticity of a Ceramic Body

Many years ago during a conference in Germany, W.D. Kingery drew my attention to a small book entitled "Clay Water Systems" published in 1965 by W.G. Lawrence of Alfred University. Even today I still consider this publication as one of the best sources on the subject of plasticity. Amongst other topics, it deals with interesting tests relating to the factors that determine the plasticity of clay, respectively, ceramics. Of course, optimization of plasticity confronts us with economic, technical and material-based limitations. However, it does make sense to keep in mind where the adjustment screws are! I do not intend to get lost in the different factors, but have produced a small illustration that might be worth referring to should you encounter problems with the plasticity of your ceramic body (Fig. 6.4).

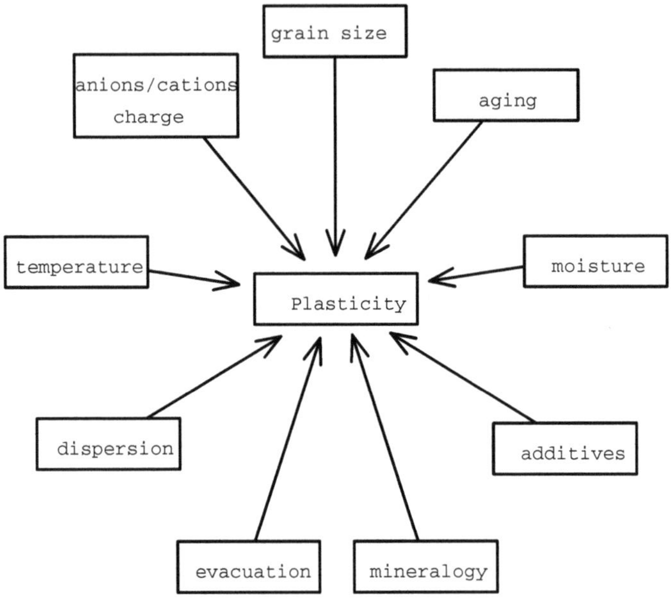

Fig. 6.4 Factors determining plasticity of a ceramic body

6.5 Two Methods to Measure the Development of Plasticity

Let me now introduce two useful methods.

6.5.1 Good-Natured and Nasty Ceramic Bodies—the Consistency Curve

In one of my studies, I determined the dependency between the consistency—I will come back later to this term—and the water content of six European clays for brick-making. For these tests, I used the so-called Brabender Plastograph, a type of small kneader. The various clays were kneaded at different water contents between 0 and 30%. After a kneading time of about 5 min for each water content, the detected torque of the kneader was measured. This torque gives an idea of the consistency of the mass, or in other words, of the toughness of the mass. The higher the torque, the tougher the kneaded mass. The first relevant finding to remember is that there are ceramic bodies that I tend to call "good-natured" meaning that minor changes in the moisture content will not lead to relevant changes with respect to the consistency of these bodies in Fig. 6.5 curve A and B. On the other hand, you can find materials that show downright "nasty" behaviour. The slightest variations in the water content result in dramatic changes in the consistency, in Fig. 6.5 curve C. The second important finding from these tests is that a clearly distinctive maximum reading for all ceramic bodies can be determined: quite high for curve A, less high for B and C.

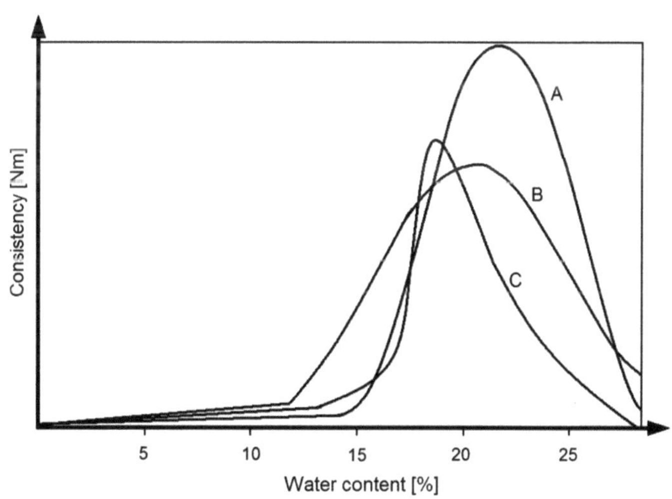

Fig. 6.5 Development of consistency in relation to water content

As this characteristic can be established with little effort for different ceramic bodies and because of the paramount relevance regarding the extrusion process, I consider it a must to know the "consistency curve" in particular for those bodies which I have called nasty.

6.5.2 *Kneading Curve*

A second method had to be developed when we started to concentrate on advanced ceramics where the two meanings of plasticity are even more relevant if one thinks of honeycombs with walls of 0.2 mm thickness which should not deform after extrusion. We call it the "kneading curve"—see Fig. 6.7. Here we look into the question of how we can change the extrudability of a ceramic mass by kneading—that is by the introduction of shearing energy into the ceramic body at a constant moisture content. In order to ascertain this, we have designed an instrument called the Capillary Check (Fig. 6.6).

Fig. 6.6 Capillary Check

With the Capillary Check, we are able to detect the relevant rheological parameters such as radial and axial pressure, wall slippage between the ceramic mass and the surface of the die as well as the influence of the so-called shaping degree and the shear stress, etc. The Capillary Check is a sort of rheometer with which we can simulate extrusion. On the graph—see Fig. 6.7—you see three types of ceramic bodies.

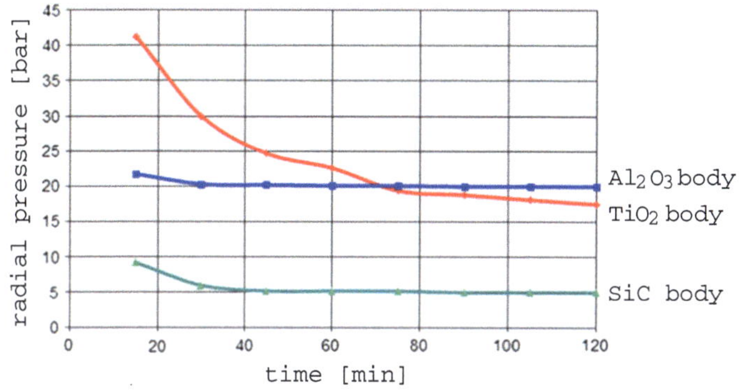

Fig. 6.7 Kneading curve for three ceramic bodies

The *x*-axis shows the kneading time or the introduced shear energy expressed in kWh; the *y*-axis shows the pressure rate measured at the die of the Capillary Check. How does it work? The ceramic body is mixed in a batch-type kneader. A sample is collected every 5 min for as long as the pressure rate becomes asymptotic; typical for all ceramic bodies. The point where the curve is becoming asymptotic, we named it "optimal extrusion point". The meaning is that the kneading time or shearing energy for the extrudability of this specific body reached its optimum. Again we believe it is a must to know the "kneading curve" for your bodies, whatever will be the conclusions.

6.6 Plasticity or Consistency or Extrudability?

In the last chapter, you suddenly discovered the terms consistency and extrudability and maybe you are confused; but you are not alone! Let us explore how Brownell in his book "Structural Ceramics" defines these terms:

> Consistency of a clay-water system is simply its resistance to shearing forces such as in stirring. Of course, it is greatly dependent on water content. Any clay has low consistency at very high and very low water contents. Presently we are going to see that consistency goes through a maximum as the water content is increased from zero to excessively high values.—See Fig. 6.5.

> Extrudability is a term often used in the industry where extrusion is the method of forming, and it refers to the ease of extruding a clay mixture through a die. It is not closely related to consistency, because a clay material will extrude easily at maximum consistency, but if the consistency is lowered by the removal of water the extrudability may become very difficult. On the other hand, if the maximum consistency is lowered by the addition of water the ease of extrusion increases.

Can we call the values detected on the consistency curve and the kneading curve as plasticity? Of course not! Plasticity I and II, as we have defined them, are material-specific items of a green plastic ceramic extrudate. On the other hand, the pressures detected in the kneading curve or the consistency rates we measure are process-relevant parameters of great importance for the extrusion process. Obviously, these readings depend on the plasticity of the ceramic mass, but not only on that. A helpful term in my opinion is extrudability which basically depends on at least four relevant factors: plasticity of the body, shaping degree, column speed and wall slippage in the system; but you know this already!

6.7 Conclusions

- Assuming you want to know if your material, respectively your product, can be extruded and how it can be extruded, the easiest way to find out is a simple test with the Capillary Check or—more extensively—with a suitable auger extruder or piston press.
- On the samples, detect the deformation pressure and gain an idea regarding the rupture strength.
- If you want to know if the extrudability and/or plasticity of your body can be optimized by intensive kneading, you should establish the kneading curve of your mass.

Bibliography

1. W.E. Brownell, *Structural clay products* (Springer Verlag, New York, 1970).
2. T. Haase, Die Bildsamkeit plastischer keramischer Massen. Berichte der DKG **34**, 27–33 (1957).
3. W.G. Lawrence (ed.), *Clay Water Systems* (Alfred University, New York, 1965).
4. F. Moore, *Rheology of Ceramic Systems* (Maclaren & sons, London, 1965).
5. P.C.F. Møller, et al.,Yield stress and thixotropy: on the difficulty of measuring yield stresses in practice. Soft Matter. Roy. Soc. Chem. (2006).

Chapter 7
Mixing and Kneading

As long as the cake tastes so good, I don't care about the homogeneity of the dough.
—My favorite baker

Abstract The principles and differences between mixing and kneading with regard to the extrudability of the ceramic body are outlined as well as the most relevant terms like dispersion, micro- and macro-mixing. Mixing or kneading ahead of extrusion and following crushing and dosing is a complex process consisting of various elements like continuous or discontinuous process systems which can be arranged in different configurations to achieve the specific requirements for an optimal extrusion.

Keywords Mixing · Kneading · Shearing · Macro-/micro-mixing · Dispersion · Buffer systems

7.1 Short Introduction

Extrusion in the ceramic industry assumes the mixing and/or kneading of the plastic material after the dosing stage, as it is extremely rare that the so-called feedstock is available in a homogenized and plasticized form; it rather consists of several granular and frequently pulverized and liquid components, which must be formed to a plastic body. An adequate constant homogeneity and plasticity are an indispensable prerequisite for extrusion.

7.2 Mixing

> The aim of mixing is to achieve a uniform distribution of all mix components to reach a more or less homogeneous system. Schubert—see references in this chapter.

F. Händle, *The Art of Ceramic Extrusion*, https://doi.org/10.1007/978-3-031-63471-0_7

This quote from Schubert makes it sufficiently clear what is meant by mixing. The theoretical principles of mixing are not to be discussed here in greater detail. Good contributions to the theory of mixing are easily accessible on the Internet. That mixing can be performed either well or not so well sounds trivial, yet in practice, this can often not be ascertained quite so easily. Although there are existing methods, such as the fluorescence process, to determine the so-called mixing quality, even experienced practitioners are not completely convinced of this, at least in ceramics. Therefore, as an alternative, the mixing quality is frequently measured in a quasi-secondary way using dried specimens and determining their bending strength. However, it makes sense to differentiate between four mixing conditions: complete de-mixing, ideal distribution, textured mixing and stochastic distribution (Fig. 7.1).

In practice, mixing in ceramics as binary system stochastically distributed rarely exists, rather than multi-component systems consisting of different solids and liquids.

The mixing and plasticizing processes are often accomplished in several steps. In brick factories, for example, after the raw material has been premixed during the preparation process, it is transferred to souring basins not only for secondary mixing, but also for simultaneous ageing, as many ceramic materials benefit from increased plasticity as a result of the ageing process. Basically, we differentiate

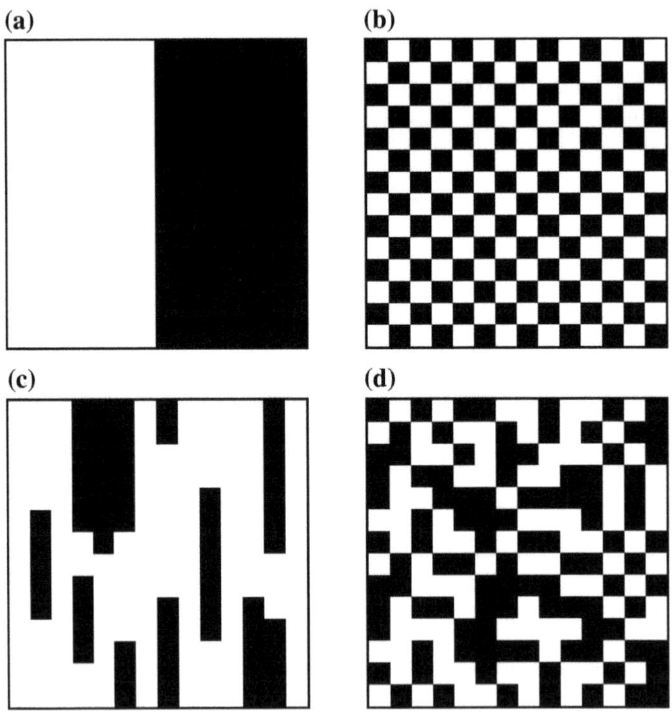

Fig. 7.1 Mixing conditions **a** complete de-mixing, **b** ideal homogenous mixing, **c** textured mixing, **d** stochastic homogenous mixing

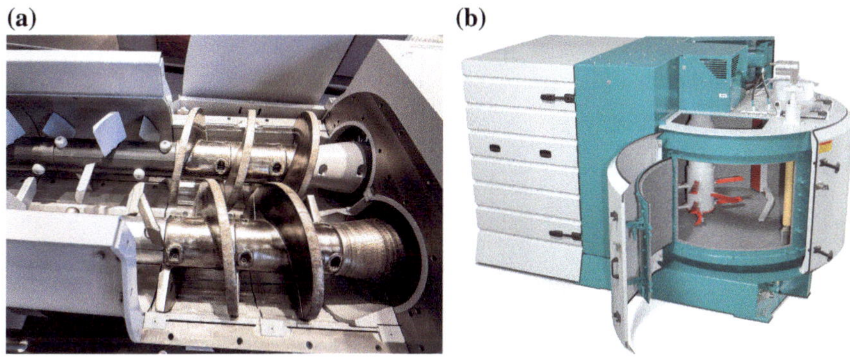

Fig. 7.2 Continuous (Händle GmbH) (**a**) and discontinuous (Eirich GmbH & Co KG) (**b**) mixers

between continuous and discontinuous mixers of different designs. Figure 7.2 shows prototypically pictures of an intermittent and of a continuous mixer.

7.3 Kneading

If you hear the word kneading and you associate this spontaneously with the kneading of a cake mixture or pizza dough, you are spot on. In the case of good chocolate, this process is accomplished by conching between rollers. The quality of many food products depends on intensive kneading, and the same applies also to such materials as natural rubber or hard metal, whereby kneading is primarily the introduction of shearing energy. Similar to mixing, we can differentiate between continuous and intermittent kneaders and here again a great deal of different designs is available (Fig. 7.3).

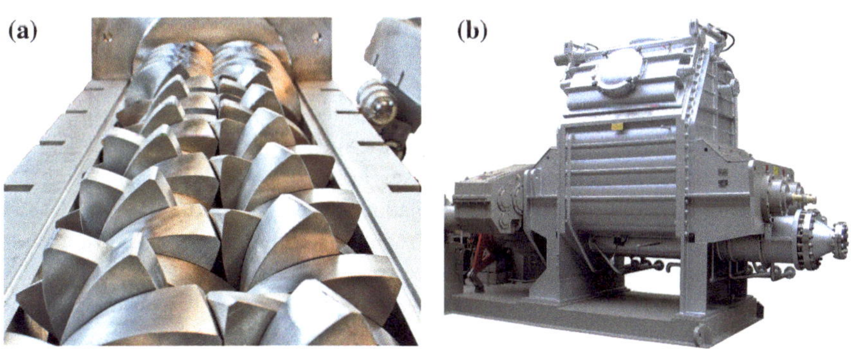

Fig. 7.3 Continuous (Hasler + Co AG) (**a**) and discontinuous (AMK GmbH & Co KG) (**b**) kneaders

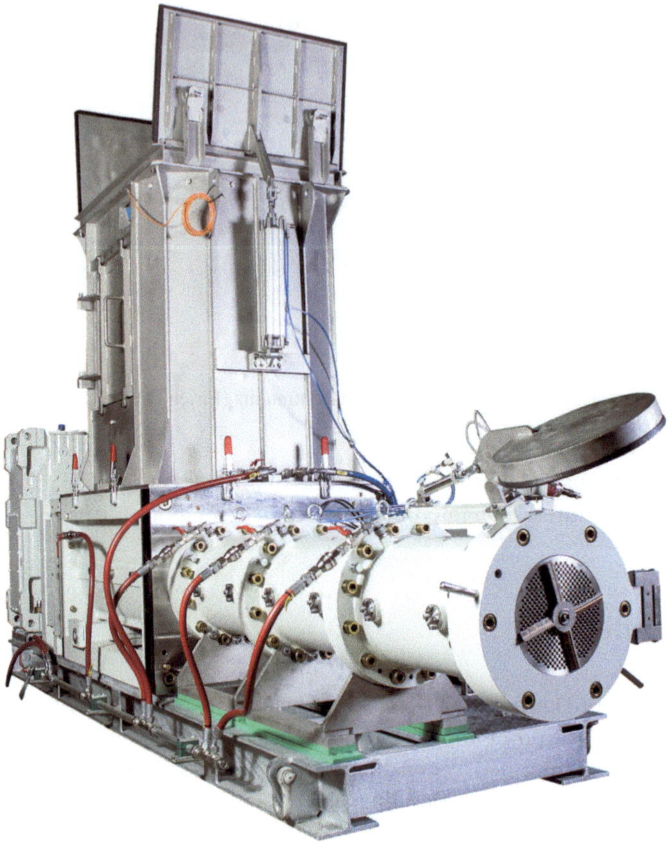

Fig. 7.4 Dosing strainer (ECT KEMA GmbH)

Where does the difference between mixing and kneading lie in the ceramic industry, bearing in mind that many companies for various reasons use intermittent kneaders for mixing; thus, mixing and kneading are effected in parallel. Those supporting intensive-type mixers might rightly comment at this point that these intensive mixers provide mixing very effectively combined with an added shearing effect. In most cases as confirmed by numerous tests, the shearing energy generated in intensive mixers is indeed sufficient. In certain cases, however, an additional shearing, that is to say kneading, in either intermittent Z-blade kneaders, or continuous kneaders, or extruders with extra kneading elements, is required. With the help of the "kneading curve"—see Chap. 6—this statement can be easily and empirically proved.

A great deal of the know-how of manufacturers of high-quality extruded ceramic products lies hidden in this process. Even though the generation of the required degree of plasticity or extrudability in producing filigree extrudates such as honeycombs is of such paramount importance, there are hardly any references in the literature to the fact that certain flow characteristics of ceramic bodies can only be achieved by introducing high shear energy.

7.4 Macro-mixing and Micro-mixing

In the literature, you often find a distinction between macro- and micro-mixing, respectively, convection mixing versus diffusion mixing. I will try to explain this without having to burden you with terms from so-called tribochemistry such as mechano-technical bonding, van der Waals forces, etc. Macro-mixing assumes that the larger particle collectives change their position in the composite during the mixing process; it is for this reason that we also speak of transposition processes. In contrast to this, micro-mixing features a transposition of the individual component within the particle collective caused by the influence of the shear forces. For our purposes, this definition is necessary though not quite adequate because micro-mixing assumes the existence of individual particles, which implies a de-agglomeration of the particle collective to individual components. Could this explain the effects with certain bodies that despite intensive mixing, higher plasticity readings cannot be reached? Or does the effect of kneading by the introduction of high shear forces result primarily in a de-agglomeration of individual components? I will come back to this.

Many tests have demonstrated that conventional mixers are often not quite sufficient to reach an adequate level of plasticizing since the particle collectives can evade during the mixing process. In this case, longer mixing does not result in a material conversion but would only amount to an increased temperature of the mix. A similar phenomenon can also be found with conventional Z-blade mixers, where the material cannot be forced into the shearing gap due to the geometry of the kneading tools.

7.5 Buffer Systems or Dosing Strainers?

Just to repeat and summarize the mixing and kneading options. For each case, we have four options in distinct variations: continuous and/or discontinuous mixers versus continuous and/or discontinuous kneaders. Discontinuous mixing or kneading followed by continuous extrusion necessitates storage buffer systems between mixing or kneading and extrusion. Buffer systems in which preferably no segregation or drying of the body should take place.

There exist proven buffer systems like box feeders or silos; however, these cannot meet the requirements for sophisticated ceramic products such as honeycombs, filters and fuel cells. For this reason, the so-called dosing strainers of different sizes have been developed over recent years. These are in fact multifunctional systems for buffering, homogenizing and dosing, and also, on request, for temperature control and filtering (Fig. 7.4).

7.6 The Ceramic Body as Dispersed Material System

In order to understand what happens in plasticizing of certain ceramic bodies, we cannot in my opinion ignore what the term dispersing means. The dispersed material system consists of several or at least two phases, one being the dispersed or inner phase which is surrounded by an outer continuous dispersion phase. The elements of the dispersed phases are always discreet particles regardless of their physical state for instance grains, droplets, bubbles, etc. Since the size of the particles is highly relevant within the framework of material conversion, one differentiates between:

- molecular dispersed material systems (<1 nm)
- colloidally dispersed material systems (>1 nm, <1 μm)
- coarsely dispersed material systems (>1 μm).

Whereas the interaction forces between the particles can be neglected for coarsely dispersed material systems, the material conversion of colloidal dispersed and molecular dispersed material systems is affected by the introduction of chemical reactions. In the ceramics industry, different machines are employed for dispersion, mainly kneaders, three roll mills, shear roll extruders and extruders. There are a few fundamental characteristics that differentiate the disperse systems usually found in the ceramics industry from ideal disperse systems:

- Multi-component systems which may consist of different ceramic minerals, inert and organic components and, in the liquid phase, besides water also of liquid additives, dispersing agents, lubricants such as olein or glycerine, methyl cellulose.
- Systems in which the particle size of the solid phase show a grain size ranging from coarsely to molecular.
- Depending on whether the particles are dual-layer or single-layer minerals, the liquid phase can be embedded in the crystal lattice and thus accordingly changes the degree of de-agglomeration during dispersion.

Does this general representation of dispersion help us to explain whether or what happens for instance during the kneading of ceramic materials? It does not actually help us further if we ascertain that these are highly complex processes, even if that is true. What we really need are instruments and methods to determine whether the consistency of the ceramic body is good enough for extrusion or whether this can still be optimized by extra kneading.

7.7 Conclusions

- Mixing and kneading are two different operations: mixing is primarily the stochastic distribution of particle collectives; kneading is primarily the de-agglomeration and dispersion of particles and particle collectives by shearing.

- Both processes overlap during mixing as well during kneading.
- Ceramic bodies may be understood as coarsely to colloidally disperse multi-component systems.
- Try to analyse exactly which of these options is the optimal solution for you.
- Find a target figure and method to prove and evaluate if your body has the right homogeneity for the extrusion process.

Bibliography

1. E. Blass, *Entwicklung verfahrenstechnischer Prozesse* (Springer Verlag, Berlin Heidelberg, 1997).
2. L.P. Edward, et al. (eds.), *Handbook of Industrial Mixing: Science and Practice* (Wiley, New York, 2003).
3. G. Heinicke et al., *Tribochemistry* (Akademie-Verlag, Berlin, 1984).
4. M. Kraume (ed.), *Mischen und Rühren* (Wiley-VCH Verlag, Weinheim, 2003).
5. S.M. Kresta, et al. (eds.), *Advances in Industrial Mixing: A Companion to the Handbook of Industrial Mixing* (Wiley, New York, 2015).
6. ChG Schofield, *Homogenisation/Blending Systems Design and Control for Minerals Processing* (Trans Tech Publications, Rockport, MA, 1980).
7. H. Schubert, *Aufbereitung fester mineralischer Rohstoffe* (VEB deutscher Verlag für Grundstoffindustrie, Leipzig, 1989).
8. R. Weinekötter et al., *Mixing of Solids* (Springer Verlag, Heidelberg Berlin, 2000).
9. N. Harnby, M.F. Edwards, A.W. Nienow (eds.), *Mixing in the Process Industries* (Butterworths, London, 1985).
10. E.L. Paul, et al. (eds.), *Handbook of Industrial Mixing: Science and Practice* (Wiley-Interscience, New York, 2003).

Chapter 8
Extrusion as a System

Only variety absorbs variety.
—Ross Ashby

Abstract Extrusion can be considered as a subsystem within a specific manufacturing system for ceramic products. The benefits of a system-oriented approach have been widely accepted in many scientific disciplines as an essential element for situation analysis and problem-solving.

Keywords System theory · Ceramic production system · Feedback · Interdependencies between subsystems · System attributes

8.1 What Do We Understand by a System?

Word has spread of late that system-oriented approaches and methods make sense, or are even absolutely essential, in describing complex structures and processes as part of problem analysis and solving. Moreover, complex problems can only be solved by using complex approaches. Over recent decades, system theories of different orientations have been developed in applied sciences such as sociology and biology, information theory and engineering, all of which, however, are more or less based on the same theoretical fundus. Part of this theoretical fundus—the general system theory—is the groundbreaking "Law of Requisite Variety" described by Ross Ashby in his book "An Introduction to Cybernetics" (1956). Put simply, the message of this law is: "Only complexity absorbs complexity".

I am well aware of the fact that the popular recommendation of reducing complexity is a mantra in management training: in certain cases absolutely good advice. This does not make sense, however, whenever the complexity of the situation or problem cannot be reduced without loss of reality and information. In the framework of this chapter, I will try to provide an abridged insight into some elements of system theory and to explain, using practical examples, the usefulness of system-oriented approaches in solving problems in ceramic extrusion.

F. Händle, *The Art of Ceramic Extrusion*, https://doi.org/10.1007/978-3-031-63471-0_8

Reading the following definitions and explanations, you might question the usefulness of such terms in our context. Please understand it as a little training to sharpen your insight into the behaviour of complex technological systems with which you are confronted anyway; whether you like it or not.

8.2 Components of a System

The system "ceramic production" can be understood as a target-oriented system meaning a structure or process which on the one hand can be differentiated from its surroundings and which on the other consists of subsystems—such as the subsystem extrusion—whereby the individual subsystems can be finely broken down to the *n*th degree into elements or subsystems (Fig. 8.1).

This defines the first component of system theory whereby the definition of a system and its units always implies the definition of limits to the superordinated, coordinated and subordinated units. It depends entirely on our interest respectively target of the problem analysis, to how finely we break down a system into the vertical and horizontal planes. If we consider a system as a process, the subprocesses allow themselves to be arranged in a temporary configuration. For example, preceding extrusion upstream is the dosing and preparation process of the ceramic body, whereas the following extrusion downstream are handling, drying and firing of the products. We can assume that the extrusion process depends stronger upon the upstream processes than upon the downstream processes. It should be noted that with every production system, we are dealing with two inseparably interlinked systems—a social system and a technical system. We speak therefore of a socio-technological system. Technical problems are often found to be in fact social problems: for example contamination of the body as a consequence of inadequate maintenance or, vice versa, as a consequence of excessive demands on the workforce caused by non-transparent technical

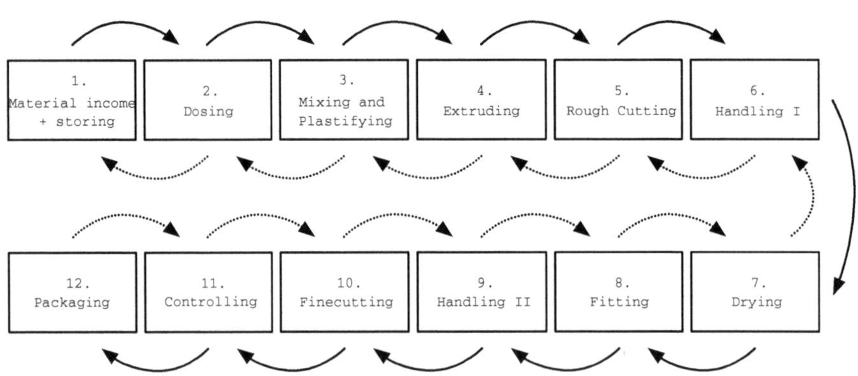

Fig. 8.1 Ceramic production system

complexity. Communicative, energetic and material relationships of different quantity and quality exist between the individual units of a system, its subsystems and components; these are called linkages, dependencies, relationships and interdependencies. For instance, the result after firing depends on a correct drying process or if the preparation system does not provide material in sufficient quantity, the extruder cannot produce the specified throughput rate. It is important to quantify, qualify and evaluate these dependencies as accurately as possible. An important type of interlinking is the so-called feedback. This is according to Wikipedia a phenomenon in "information-processing systems whereby a part of the output value is fed back directly or in modified form to the input of the system". A distinction is made between positive and negative feedback.

An example of negative feedback: let us assume that the viscosity of a ceramic body is decreasing. As a result, we face a backlog in the feeding zone of the extruder and a reduction of the default column speed which is controlled by a sensor. This in turn leads to an increased revolution of the auger to comply with the specified column speed which in a second turn leads to another decrease of the viscosity because of the extra shearing of the ceramic body and so on.

Interlinking and feedback between systems, subsystems and components can occur in the form of linear and nonlinear, dynamic or static, deterministic or stochastic, simple or complex, continuous or discrete, non-simultaneous or simultaneous coupled dependencies, whereby the different interlinks may be superimposed, as can be observed with such complex systems as the weather or with ecological systems.

Finally, one can differentiate between learning systems and non-learning "dumb" systems, which are already of great significance today and which in the future will be of immense significance in view of the rapidity of developments; just think of such keywords as "Internet of Things" or "Artificial Intelligence", Self-organization, etc. In Table 8.1, you will find a list—not at all complete—which shows some of the attributes of the different systems.

8.3 Consequences

For the production of high-quality ceramic items such as honeycombs, complex filters or other high-grade products, it makes sense to consider extrusion as a subsystem/process within the complex production system/process. Any deviations or intervention in the process steps upstream of the extruder will result in deviations during extrusion, which in turn can cause deviations in the drying, firing, etc. Conversely, any requirements in respect of the final product in terms of quality and quantity such as adherence to tight tolerances, assume certain measures for extrusion which must be checked for their feasibility along with the specified basic conditions like the formula and rheology of the body. Unacceptable deviations on the one hand and any unrealistic requirements on the other are therefore just two sides of the same coin

Table 8.1 System attributes

Attribute	Predication		
Existential realm	Ideal	Real	
Manner of origin	Natural	Artificial	
Environmental connection	Isolated	Closed	Open
Time dependence	Static	Dynamic	
Temporal attitude of attributive values	Continuous	Discrete	
Type of function	Linear	Nonlinear	
Deg. of determination	Deterministic	Stochastic	
Type of relations	Unilateral	Reciprocal	
No. of subsystems	Few	Numerous	
No. of relations	Simple	Complex	Extremely complex
Deg. of hierarchy	Hardly hierarchical	Very hierarchical	
Type of transfer	Material	Information (data)	Energy
Type of system	Subjective	Active	Objective
Objective orientation	Nonpurposive	Purposive	Functional
Behaviour	According to the principles of primary regulation	According to the principles of secondary regulation	
Activity level of relations	Active	Inactive	

and must be analysed and possibly modified "top-down" and "bottom-up" along the entire process chain.

Here is a recent case in point: a small tube should be extruded from an obviously intrinsically viscous ceramic material that expands after extrusion. Outside diameter of the sintered tube is 2.3 mm, the inside diameter is 1.9 mm and it has a length of 12 mm. The admissible tolerance is supposed to be plus/minus 3/100 mm. Is this feasible? Instead of critically questioning such specifications, the situation in practice is often different. When defects occur in the drying process, the root of the problem is frequently searched for in the upstream extrusion process, not in the drying as such, nor in the process steps ahead of extrusion, nor in the specifications which perhaps may be unrealistic. This shortened, simplistic method at a preliminary stage is frequently the reason that projects are not successful and there is a failure in problem analysis.

Most of the problems that appear in extrusion in the form of quality and/or quantity problems, or are traced back to extrusion, in fact, have causes "outside" extrusion. The most common external causes ahead of extrusion are variations in inadequate preparation resulting in inhomogeneity or lack of plasticity. For example a too fast drying process or inappropriate handling causes deformation of the extrudates. Internal causes for extrusion problems can lie in the failure of subsystems of the extrusion system: auger wear, malfunction of sensors, insufficient temperature control, errors in de-airing—see also Chap. 13.

External and internal causes are often superimposed and complicate a complex problem analysis. Whenever deviations cannot be avoided, be they caused by seasonal variations in the moisture content of the raw material, or variations due to climatic conditions, it is recommended that a catalogue be compiled detailing which of the deviations in subsystem A causes which deviations in the remaining subsystems B-X. Secondly, the admissible corridors for the primary and the derived secondary deviations should be defined, i.e. the "pain thresholds". Thirdly, a catalogue of measures should be prepared to record whether and how certain deviations could be compensated.

Let us assume a body that is extremely drying sensitive and prone to laminations and from which filigree extrudates should be made. This raises two questions: how, on the one hand, can the individual process the steps to be optimized in view of these constraints, e.g. intensive kneading and improved flow conditions in the pressure head and die in an effort to minimize potential laminations, etc. Generally speaking, the question is how can the downstream processes be optimized by changes within the upstream process steps. On the other hand, do we have to accept the boundary conditions relating to the susceptibility to laminations and drying sensitivity of the body as given?

Here is another example: let us assume that transport and handling cause deformations in the green extrudates. Again it should be attempted to approach the problem from two different directions, directly by optimizing the transport technique and indirectly by increasing a higher deformation pressure of the body.

In the words of the economist Joseph Schumpeter:

> We often have at our disposal a much larger "box of tools" to enable us to repair any deviations and, at the risk of becoming unpopular, we should have the courage to question specifications which seem to be "cast in stone".

For deviations in extrusion, system theory-oriented methods enable us to analyse, evaluate and eliminate or reduce the external and internal causes. Restricting ourselves to internal causes rarely leads to success, as is the case if we concentrate on external causes, which often results in moving the responsibility backwards and forwards between the suppliers of the different subsystems, especially if the interfaces were not clearly defined.

8.4 Conclusions

- Familiarize yourself with practice-oriented systemic approaches which can be of great help in analysing and solving complex problems in extrusion.

- Attempt to find the causes for extrusion problems both internally and externally, i.e. also in the subsystems of the production systems upstream and downstream of extrusion. Precisely define the limits of the analysed system and those of the individual subsystems and elements.
- Compile a catalogue of measures recording which dependencies exist between extrusion and the other external subsystems as well as internal elements, detailing those deviations where these result in deviations in extrusion, plus how and by what actions these can be compensated.

Bibliography

1. W.R. Ashby, *An Introduction to Cybernetics* (Chapman & Hall, London, 1956).
2. P. Checkland, *Systems Thinking, Systems Practice* (Wiley, New York, 1999).
3. W. Churchman, *The System Approach* (Dell Pub, New York, 1984).
4. P. Claussen, *Die Fabrik als soziales System* (Springer Gabler, Wiesbaden, 2009).
5. W.F. Daenzer, F. Huber (eds.), *Systems Engineering* (Verlag Industrielle Organisation, Zürich, 2002).
6. J.W. Forrester, *Principles of Systems* (Pegasus Communications, Waltham, 2000).
7. F. Händle, The Brickyard as a System, in *Brick and Tilemaking*, ed. by W. Bender, F. Händle (Bauverlag, Wiesbaden, 1982).
8. A. Pickering, *The Cybernetic Brain* (The university Chicago Press, Chicago, 2011).
9. D. Sherwood, *Seeing the Forest for the Trees* (Nicholas Brealey Publishing, London, 2002).
10. J. Sterman, *Systems Thinking and Modelling for a Complex* (World. McGraw-Hill, New York, 2000).

Chapter 9
Laminations and Chaos

Without order nothing can exist—without chaos nothing can evolve.
—Albert Einstein

Abstract The term lamination referring to extrusion is defined. The formation of laminations is explained, and the different types of laminations are described. On this basis, suggestions on how to avoid laminations are proposed.

Keywords Laminations · Causes and formation of laminations · Extrusion laminations · Types of laminations

9.1 What Do We Understand by Laminations?

For our purposes here we shall use the term laminations, although I am aware of the fact that laminations can also be understood in the sense of patterns. Below is a limited definition of the phenomenon of lamination within our frame of reference.

We understand laminations to encompass faults within the structure of the extrudate which have stemmed from various causes and which predominantly result in quality problems. The origin of these disturbances could be in the extrusion process as such, or in the subprocess upstream or downstream of the extrusion; also in the ceramic body itself. It is mostly the case that a combination of various causes comes together. In such cases, we rightly speak of "overdetermination". In practise, it is difficult to isolate the individual sources.

In a joint article written together with Rainer Bartusch, we differentiate between three basic types of variables: ceramic variables, process-engineering variables and mechanical processing variables (Table 9.1).

An example: if the body has not been mixed sufficiently in the preparation stage or has not been dried adequately after extrusion, this can lead to all kinds of laminations that are not attributable to extrusion. The causes of these laminations are only mentioned in passing; the focus of the following explanations is those specific laminations typical for shaping with auger extruders. Incidentally, laminations can be

F. Händle, *The Art of Ceramic Extrusion*, https://doi.org/10.1007/978-3-031-63471-0_9

Table 9.1 Causes for laminations in extrusion

Ceramic variables	Process-engineering variables	Mechanical processing variables
• Type of raw material • Mineralogical constitution • Ion arrangement • Water content/body stiffness • Opening agents • Other additives	• Preparation/homogenization • Body or column temperature • De-airing • Screw shaft speed • Column progress • Column rate • Oiling or moisturizing of the extruder head and/or extrusion die • Intermediate transport and moist storage • Drying • Firing • Aggregates	• Geometry of screw • Pointed screw head • Cylinder casing • Extrusion head • Extruder die

experienced with all kinds of ceramic shaping processes—see also Chap. 2. Anyone who claims that there is a completely free lamination shaping method in ceramics has—in my opinion—not looked carefully enough at the specific laminationsof the different shaping methods. But there is no question about the specific and occasionally difficult boundary condition for extrusion: the shrinkage of the extrudate during drying as a result of the higher moisture content of the body.

In the following, we will try to differentiate various aspects of the subject:

- How are laminations generated?
- What type of laminations do we face within extrusion?
- Which of these laminations are extrusion-specific and do not occur with other shaping processes?
- Can these laminations be eliminated or minimized?

9.2 Formation of Laminations

Based on our definition that laminations are "faults within the structure of the extrudate", I would mention at this point that the following explanations do not deal with laminations caused by inadequate or variable grain structure of the body or laminations resulting from insufficient crushing, grinding, mixing or drying. There is a substantial amount of literature available on the subject of laminations in extrusion, also imaginative nostrums for so-called lamination destroyers of all kinds. I shall limit myself here to two fundamental works dealing with the subject of the formation of laminations.

9.2.1 *The Pioneering Studies by BCRA*

The studies by H.R. Hodgkinson on behalf of the British Ceramic Research Association (BCRA) in 1962 and the various contributions by Pels Leusden, which primarily appeared in different publications between 1970 and 1990, are fundamental for our topic. I still admire the works and the supporting educational film made by the BCRA in the 1960s under what we today would consider the most primitive conditions. They were truly groundbreaking in understanding the flow characteristics of ceramic masses during extrusion, both from an empirical and theoretical point of view.

These works—see Fig. 9.1—basically demonstrate the flow pattern of ceramic bodies through dies of different geometries. Changes were made to the exit width of the dies and the resulting exit angles. There is a clearly visible dependence between the body, which runs faster in the centre of the column, and the exit angle. The steeper the angle, the faster the column speed in the centre. Moreover, the pictures clearly show that a too-steep angle creates a risk of "dead corners" where the body does not move at all and may therefore dry out. Slip planes create varying flow rates within the column, the consequence of which can be the formation of laminations. These are the so-called flow laminations. In my opinion, there exists another type of flow laminations which I call flow laminations II—more on that later.

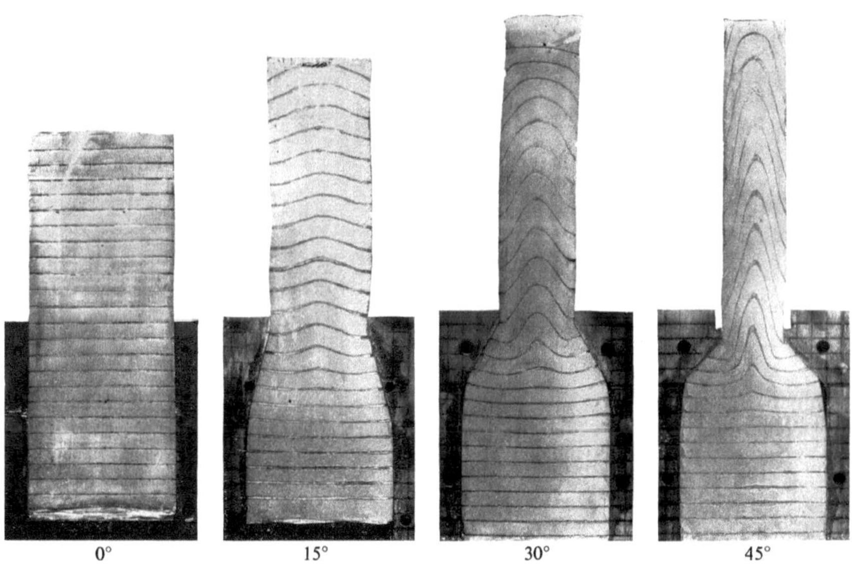

Fig. 9.1 Effect of the geometry of die upon lamination

9.2.2 *Where Are Laminations Generated in Extrusion?*

In his works, Pels Leusden was able to define where laminations are generated in the auger extruder by using an ingenious experimental arrangement.

It can be clearly seen in Fig. 9.2 that the level of lamination is lowered in the pressure head, whereas it increases to its highest level in the die, where the actual shaping process takes place. It therefore makes sense to concentrate on pressure head and die when considering measures as to how laminations can be reduced, however, without neglecting those laminations generated by the auger of the extruder. It is understandable that models of the chaos theory were used to illustrate and simulate the flow patterns of ceramic bodies within pressure head and die.

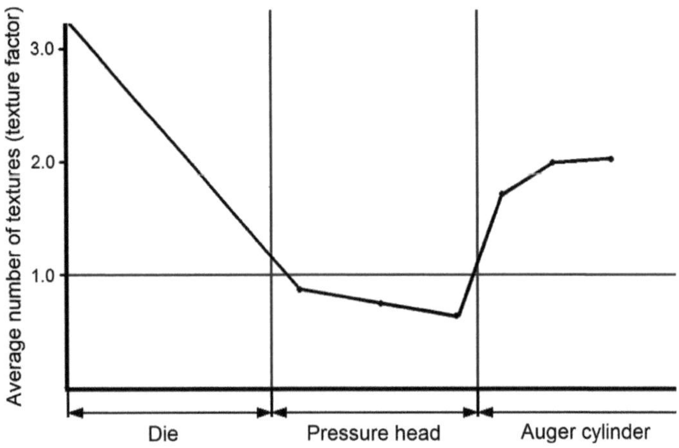

Fig. 9.2 Generation of laminations in the auger extruder

9.3 Types of Laminations

In light of the complex processes during the shaping of ceramic bodies when extruded, it makes sense to analyse which type of laminations are involved. Especially as we know that, in practise, different types of laminations overlap. In the following, only those types of laminations are mentioned which are of significance in connection with extrusion:

- Flow laminations and slip planes
- Cutting laminations
- Rotational laminations
- Hub laminations
- Laminations caused by insufficient de-airing
- Combined laminations.

9.3.1 Flow Laminations

In the majority of cases, ceramic bodies are multi-component systems consisting of minerals of varied origin, grain size, grain geometry and grain size distribution, a liquid phase as with water and additives, and occasionally organic content. Looked at microscopically, the flow characteristics of such a plastic body are highly complex and chaotic. Slip planes are formed, and it comes to a concentration of ultrafine near the hub. Water may be pressed out at higher pressure rates. In the event of the auger and barrel being cooled, water film appears on the walls rand the auger results in a reduction of the wall friction. So-called drag effects of the coarser particles can be observed. The BCRA photographs clearly demonstrate how flow laminations can develop, even in those cases where we have identical cross sections between the extruder and the die, i.e. with an outlet angle of 0°.

In spite of the non-existence of any constriction by a conical die, one could see in most cases that the column advances in the middle faster and lags behind in the area close to the wall—see Fig. 9.1. This is because the inner friction of the body is less than the outer wall friction. But in the reverse case, which can be seen in Fig. 9.3, the inner friction is higher than the wall friction, and the material in those zones

Fig. 9.3 Result of lamination test with a 500-mm-piston extruder

close to the wall does advance whilst the middle does lag behind. We have named the laminations generated by the differences between the wall friction and inner friction as flow laminations II. In practical operation, both types of flow laminations are superimposed and it is therefore important to be aware of the different causes relating to flow laminations I and flow laminations II.

9.3.2 Cutting Laminations

When the body enters the die from the pressure head, the compact column is cut into several smaller columns depending on the geometry of the core bows and cores. For those ceramic bodies which have a distinct "memory", these cuts close up only after a certain distance. Bodies that are lamination-sensitive show these cutting laminations as structural breakdown in the green or dried and fired extrudate. Pels Leusden

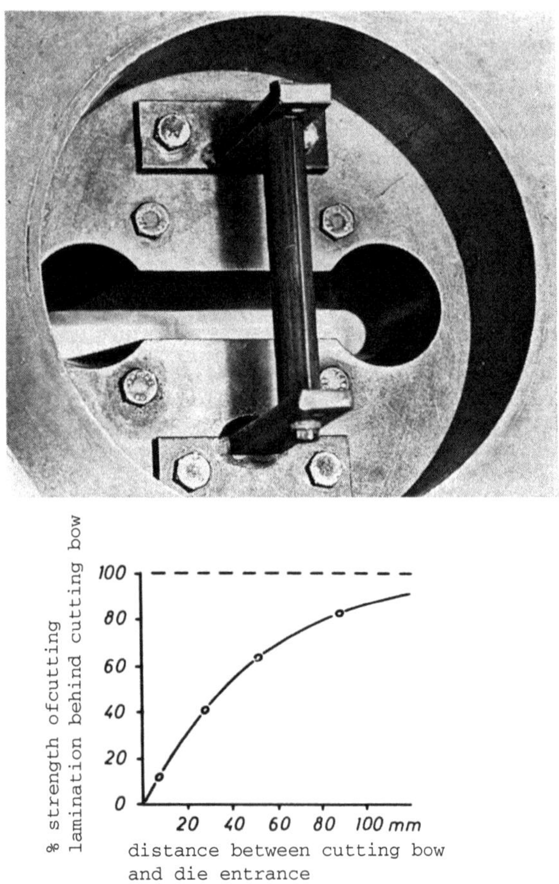

Fig. 9.4 Cutting device to test the correlation between distance behind cutting bow and original strength of the ceramic body

published some interesting works on this subject. In his contribution, he determined empirically the necessary distance behind the cutting bow to regain the original strength (Fig. 9.4).

9.3.3 Rotational Laminations

Rotational laminations occur with auger extruders firstly once the barrels are worn, i.e. the rotation of the body caused by the auger is not prevented by the grooved barrels. This effect is intensified secondly by a rough auger surface and worn diameter as a result of which the body sticks to the auger rotating with the same. From this, we can deduce how rotational laminations can be avoided. Accompanying measures to prevent rotational laminations are the use of extended pressure heads or so-called "relaxation zones" ahead of the die, as well as an optimum positioning of the double-wing top auger in relation to the auger. In contrast to flow laminations, rotational laminations are typical for single-shaft auger extruders; see Sect. 15.6.

9.3.4 Hub Laminations

When exiting the end auger, the column has a tubular shape depending on the diameter of the hub and barrel (Fig. 9.5).

Fig. 9.5 Hole in the ceramic column created by the hub

The "hole" in the tubular-shaped column can be reduced and minimized by a suitable design of the tapered end auger; but it cannot be totally eliminated. As a result of the shaping process, the column exiting the pressure head should be ideally of a compact consistency over the entire cross section. Depending on the characteristics and memory of the body, the kind of shaping process, plus the geometry of the die and pressure head, this imperfection can be found again in the dried and fired extrudate. Studies such as those of Pels Leusden point to an increased proportion of fines around the hub and, where cooled augers are applied, bodies of a higher moisture content could also be detected. Especially in the production of large honeycombs and billets for insulators, targeted measures are needed to minimize these laminations in the middle of the green extrudate, known in technical terms as "moist spots".

9.3.5 Laminations Caused by Insufficient De-airing

In extrusion not only the ceramic body is pressurized during the shaping process, but also the air contained in the body is compressed. Let us assume we extrude a ceramic body with a pressure of 50 bar. Without de-airing of the body in the vacuum chamber, say at 30 mbar, there would be an explosive reaction when the air is exposed to the atmosphere after leaving the die, equivalent to 1033 mbar. Even with inadequate de-airing, air bubbles can already be found in the body or in the form of pores in the dried and sintered extrudate. In order to eliminate these laminations, a suitable vacuum pump is needed with adequate pumping speed and sufficient ultimate pressure. But it is just as important to ensure that the body is shredded as finely as possible when entering the vacuum chamber to enable the air enclosed in the ceramic body to diffuse. It is, however, always a condition that there are no leakages in the entire extrusion system—see also Chap. 11.

9.3.6 Combined Laminations

The best-known lamination of this kind is the so-called S-crack seen in the brick industry, also appearing in slugs for insulators in the centre of a compact column; rotational laminations are superimposed by hub laminations. Caused by the rotating body, the hole in the centre develops into an S-shaped defect (Fig. 9.6).

Fig. 9.6 Typical S-crack in a quarry tile

9.4 What Can Be Done?

Can extrusion laminations be eliminated? The explanations provided above suggest that this is rather unlikely. But this is the right conclusion for a wrong question if laminations are considered as faults in the structure which cause quality problems in the finished product. Laminations, although they can be identified as such which do not result in any malfunctions, are of no interest to us in this practise-oriented context. The relevant question is therefore not whether we can eliminate laminations, but rather how we can succeed in minimizing those laminations which may result in quality or functional problems.

9.5 Conclusions

- Analyse the type of laminations. In the majority of cases, it is an overlapping of different laminations.

- Try to reduce the vulnerability of your body to laminations. This is easier said than done but quite possible. Precise dosing, mixing and kneading are also partof this, as well as are possible interventions in terms of mineralogy, through to the question of whether the effects of laminations (deformation and cracking) can be reduced by gentler drying.
- Assuming a perfect condition of auger and barrel, the optimization of the pressure head/die combination—including possible fittings such as strainer plate and "relaxation zones"—are amongst the most important tools to avoid laminations.

Bibliography

1. R. Bartusch, F. Händle, Laminations, in *Extrusion in Ceramics*, ed. by F. Händle (Springer, Berlin, 2007), pp. 187–210.
2. W. Gleissle, J. Graczyk, Rheology and Extrudability of Ceramic Compounds, in *Extrusion in Ceramics*, ed. by F. Händle (Springer, Berlin, 2009), pp. 161–172.
3. Hodgkinson HR (1963). *A Film: The Plane-Flow Extrusion of Plastic Clay*. Special Publication No. 41. The British Ceramic Research Association.
4. C.O. Pels Leusden, *Formation and Influencing of Laminations During the Shaping Process*, vol. 7 (ZI Ziegelindustrie International, 1985), pp. 390–405.
5. C.O. Pels Leusden, Die Formgebung auf Strangpressen, *Ziegeltechnisches Jahrbuch* (Bauverlag BV GmbH, Gütersloh, 1975), pp. 237–379.

Chapter 10
About the Brave New World of Sensors and Control

What cannot be measured cannot be controlled.
—Peter Drucker

Abstract The requirements relative to modern measurement and control systems for extrusion plants are described and the relevant direct production data is listed. The understanding of the mechatronic model helps to understand the function and interaction between sensors and actuators. The main reasons, as well as the relevant operating data, for the necessity of modern measuring and control systems are listed. The gain of importance by introducing so-called soft sensors is exemplary described.

Keywords Control · Mechatronic model · Sensors · Extrusion values · Soft sensors

10.1 Introduction

Modern extruders use elaborate controlling and measuring systems. The touch screen of the control system not only displays the key production data that have been measured, such as power consumption, pressure rates, temperatures or column speed. Through the aggregation of individual data, certain values can also be derived which may provide information on whether the rheology of the bodies is within the green range, whether the auger and/or cylinder are worn, etc. In the following text, I will not deal with the basics of measuring and controlling techniques, nor with programmable logic controller (PLC) or microprocessor solutions, a subject for which comprehensive technical literature is available. Instead, I will outline what is important in terms of modern control systems for present-day ceramic extrusion plants.

In my opinion, real progress in the foreseeable future relating to control and regulating technique is not as urgent as quantum leaps in the developments of sensor and measuring techniques. Perhaps you would consider it an enormous advance to control the extrusion plant from your smartphone, so what! I place greater importance on being able to measure directly and with high reliability the state of wear in different

F. Händle, *The Art of Ceramic Extrusion*, https://doi.org/10.1007/978-3-031-63471-0_10

positions of the extruder or the viscosity of the ceramic body to ensure that suitable measures can be initiated in the event of any deviations.

10.2 The Mechatronic Model

Just as a reminder, the interdependency of the mechatronic basic model is shown in Fig. 10.1. Such a system can—for instance—be presented as a block model, whereby we understand sensors as elements for the acquisition of data during a process. Actuators are understood to be those elements that, within a process control, convert electrical control signals into mechanical, pneumatic or hydraulic process correcting variables.

By online assessment, in contrast to offline assessment, we understand the continuous, qualitative and quantitative assessment of the extrusion process and of the extrudate using different measurement methods. This is based on the prerequisite that a multidimensional evaluation chart does exist, within which those evaluation criteria and corridors are determined which define whether the process and the extrudate meet the specified criteria within the defined corridors. Consequently, we can also understand an extrusion plant as a mechatronic system consisting of mechanical subsystems and elements, electronic and electrotechnical subsystems and elements,

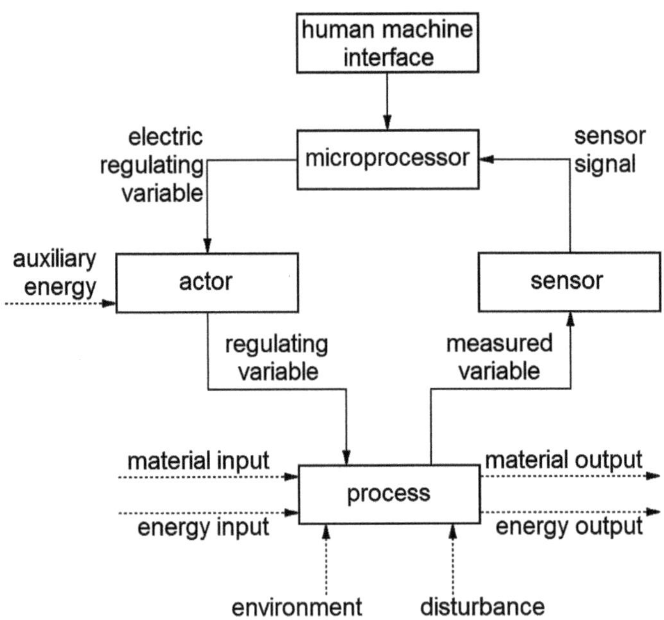

Fig. 10.1 Mechatronic basic model

as well as subsystems of IT and process data processing. There are numerous interdependencies between the overall production system and the subsystem extrusion which are interlinked to a greater or lesser degree in terms of control technology and material. The degree of networking is an important characteristic of the system; see Chap. 8.

10.3 Why There Is a Compelling Necessity for Better Measuring Systems for Ceramic Extrusion?

The key reasons for needing better measuring and controlling systems in extrusion are:

- Quality control—reduction of rejects
- Process monitoring—process reliability
- Decrease of wear—cost and maintenance saving
- Energy management—saving of energy
- Guarantee of a high utilization rate—efficiency
- Logging of the operating parameters—tracking
- Preventative maintenance—efficiency and quality control.

Sometimes, there is a huge enthusiasm for everything modern technology can offer. However do be warned against potential visual overkill, which can irritate rather than relieve the burden on the operator. Care should be taken that only those data are displayed and assessed as needed for the operation of the machine on the one hand and as are essential for the company management on the other. The two sets of data are not necessarily identical and should be selected in a pragmatic way in line with the motto "Less is more!"

10.4 What Are the Most Important Operating Data for Extrusion?

With a closed-loop ideal control system, the number of actuators corresponds to the number of sensors. This is however rarely the case with extrusion systems, since sensors are frequently only used for the recording of a measured value but do not serve as part of a control system.

Example: with most de-airing extrusion units, the level of vacuum is displayed and monitored by showing the allowed minimum—not even the maximum—reading but, although technically possible, in the rarest of cases is the vacuum regulated. Or the other way round: the rpm of the auger is measured and shown on the monitor, but for the real important value, the column speed, no sensor is used.

Table 10.1 shows the parameters which should be measured and regulated with a modern extrusion plant.

These data can be displayed on the screen in various ways and recorded via interfaces such as LAN or USB.

Already today we are in a position to determine online by way of imaging techniques if, for instance, a honeycomb shows the specified geometry, or we are able to very accurately measure the column speed online and contactless, yet there are still variables which can only be derived to greater or lesser accuracy. With the help of thermographic methods, the surface temperatures of the extruder and of the extrudate can additionally be determined, whereby ambient temperatures must be taken into account as a disturbance variable. The aspect here of primary interest is the determination of the surface temperatures, i.e. the temperature gradient over the cross section of the extrudate. Using the thermographic method for online measurement is however only suitable in very few cases. The real bad news is that up to yet there are no reliable sensors available to measure online the moisture content of prepared material upstream of extrusion. All approaches, regardless of the principle adopted, are by far too inaccurate in practice (Fig. 10.2).

Table 10.1 Measured and controlled extrusion values

Measured/controlled extrusion values	Dimension
Infeed of material in primary pug sealer	kg/s
Moisture content	%
Power consumption of primary pug sealer	amps
Power consumption of extruder	amps
Pressure rate ahead of screen plate of primary pug sealer	bar
Temperature ahead of screen plate	°C
Level control in vacuum chamber	mm
Level of vacuum in vacuum chamber	mbar
Temperature of the ceramic body in vacuum chamber	°C
Temperature of the ceramic body ahead of strainer plate	°C
Pressure rate ahead of die	bar
Temperature ahead of die	°C
Flow temperature of chiller	°C
Return temperature of chiller	°C
Column speed	mm/s
External geometry of the extrudate	mm
Internal geometry of the extrudate	mm
Surface temperature of extrudate	°C
Weight of extrudate	g/item

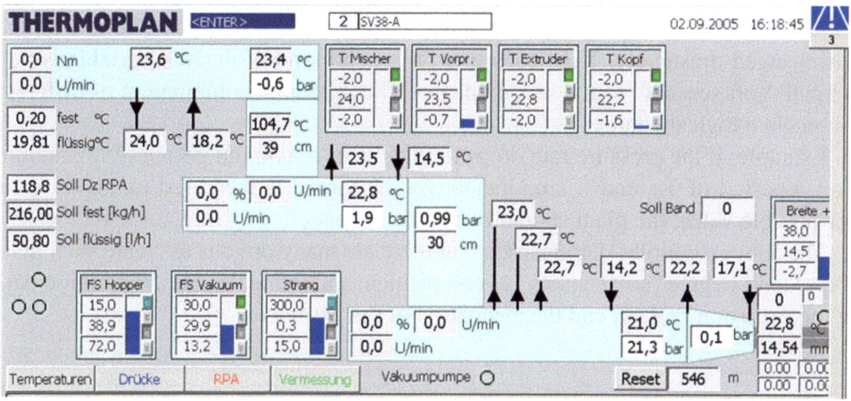

Fig. 10.2 Modern control system for an extrusion unit (THERMOPLAN GmbH)

10.5 Aggregated Measured Values/Operating Data

Apart from directly measured values, we can derive indirectly measured values, also known in extrusion technology as "soft sensors". As an example: we are aware that the throughput rate of an extruder largely depends on the wear and tear of the auger and the cylinder. That is, the same rpm of the auger is maintained but with increasing wear the throughput rate drops. Let us assume we reach a column speed of 4 m/min at an auger-shaft speed of 10 rpm for a certain extrudate, and let us also assume that this figure can only be reached at 14 rpm with a worn auger. A higher auger-shaft speed not only causes an increased temperature, whereby this could probably be compensated by an efficient cooling system. It can also result in a material backflow situation with the associated change in plasticity and finally may mean that a heavily worn auger can only be reconditioned at a high cost. It would therefore make sense to introduce a factor that tells us when an auger needs to be replaced in line with a preventative maintenance scheme. Forming the quotient between column speed and rpm of the auger relating to a certain product gives us a very simple and sufficiently reliable factor to control this problem.

Here is a second example: at a constant column speed, not only the rpm of the auger should be constant, but also the torques of the primary pug sealer and of the extruder itself should likewise be more or less constant. Let us presuppose we have a representative empirically determined factor for this aspect, and say the extruder torque should be 23,000 Nm for a column speed of 3.5 m/min. Now let us assume that this torque suddenly drops to 15,000 Nm using the predetermined column speed and material feed, controlled by a weigh-belt feeder. What has happened? The material consistency must have become softer. It is fair to say that this situation could also be verified by a reduced radial pressure. That is correct and is so in the majority of cases. However, we have found that certain operating data react to plasticity variations in a very different and sensitive manner. We have come across materials with which the pressure rate has changed only marginally in spite of lower viscosity, but the torque has changed drastically. This leads us to a second kind of derived variable which we call "soft sensors" of the second degree, these being a combination of different factors in a logical context.

Example: if the pressure rate drops to 90% of its maximum permissible pressure over a period of 5 s and if simultaneously the torque is increased to 110% of its permissible value, the plant should be stopped at once.

The above examples demonstrate that there are many options to create such first- and second-degree "soft sensors" to complement and refine the direct operation data of the extrusion process and the evaluation of the extrudate.

10.6 Conclusions

- Using the sensors available on the market plus the information-processing systems consisting of hardware and software combined with very fast and accurately operating actuators, we are nowadays able to precisely control, monitor and record online the extrusion of the most varied ceramic bodies to produce extrudates of all kinds.
- The progress achieved in recent years can be mainly attributed to the efficiency of the information-processing systems and the development of new sensors such as those used for the contactless measuring of the column speed plus imaging methods for measuring the geometry of profiles.
- Besides the directly determined measured parameters, there exists the possibility of applying "soft sensors", especially for monitoring the operating conditions which up to now cannot be measured directly.

Bibliography

1. N.S. Nise, *Control Systems Engineering* (Wiley, New York, 2015).

Chapter 11
Ancillary Equipment and Accessories

> *Ancillary: Items which are not main components of a piece of technical equipment, but do however belong to it and are used with it.*
> —Meyers Lexikon

Abstract The functioning of an extrusion system depends on adequate ancillary equipment like vacuum pump and control system. The relevant ancillary systems and devices are listed and described.

Keywords Infrastructure for extrusion · Cooling systems · Vacuum pump · Screen/die changers

11.1 The Most Important Accessories for the Extrusion Unit

Even the best de-airing extrusion unit consisting of the extruder and primary pug sealer with an intermediate de-airing chamber is not operational without accessories. In the following, we focus only on the key ancillary equipment. It is recommended to consider the following as a checklist to ensure that you have not forgotten anything.

- Foundation, base frame and supporting structure
- Control cabinet and wiring between control cabinet and motors
- Hydraulic and pneumatic units and cables
- Control, regulation and sensors
- Cooling of augers and barrels
- Dies, pressure heads and screen changers
- Vacuum pump and vacuum monitoring
- Screen or die changer
- Infrastructure.

Chapter 10 is dealing with the subject of "control" Chap. 12 with the subject "pressure head/die combination".

F. Händle, *The Art of Ceramic Extrusion*, https://doi.org/10.1007/978-3-031-63471-0_11

11.2 Foundation, Base Frame and Supporting Structure

Assuming the existence of good industrial floors, most of the extrusion plants do not require a foundation as long as the manufacturers of these plants ensure that the base frame is of such design that the permissible surface pressure for the ground is not exceeded. Particular attention must be paid to this aspect when ordering the extrusion plant, also in view of a fast installation and possible dismantling. With multi-stage de-airing units, care must be taken that the supporting structure for the primary pug sealer forms a single unit with the base frame so that all loads are transferred to the ground via the base frame. Hot-dip galvanizing or suitable coating of the base frame and supporting structure should be provided against possible corrosion and condensation at the barrels and machine body, also in view of cleaning and maintenance work. Dynamic loads can be largely neglected relating to ceramic extrusion plants; in certain cases, there might be additional loads caused by the ceramic body itself processed by the machine.

11.3 Control

Irrespective of whether the control cabinet is supplied locally or by the supplier of the extrusion plant, the same guidelines apply which however may differ from one country to another. A three-phase network of 400 V, 50 or 60 cycles, is usually assumed. The power drawn by the consumers, i.e. the motors, decides which level of amperage (A) must be available. On the basis of these conditions, the following aspects must also be taken into account:

- The size of the control cabinet depends on the power consumption of the components installed in the control cabinet.
- For higher outside temperatures and high waste heat generated by the built-in components, the control cabinet should be tempered or air-conditioned.
- In a dusty environment, the control cabinet should be protected by suitable filtration systems.
- The cross sections of the cables from the mains, respectively transformer, to the control cabinet and also from the control cabinet to the consumers must be dimensioned to suit the specific performance.
- The control cabinets must be in accordance with the respective protection classes, e.g. IP, NEMA, etc.
- The wiring within the control cabinet must be marked with different colours to comply with the statutory national provisions or the requirements of the client.
- Prior to its acceptance, all functions of the control cabinet must be checked and, in the case of certain countries, certified.
- It must be checked whether additional measures are required with respect to lightning protection and corrosion.

11.4 Cooling/Chilling of Augers and Barrels

11.4.1 Why and When Must an Extruder Be Cooled?

The plasticity and extrudability of ceramic bodies are temperature-sensitive—see Chap. 6. As a general rule, ceramic bodies can be extruded easily at low temperatures between 5 and 20 °C whilst problems may be encountered at higher temperatures. Quite often, this is not caused by the ceramic body as such but rather due to the additives used, primarily the methylcellulose binders which tend to gel at higher temperatures. Finally, it may be advisable to cool the auger, in particular, when operating at high-pressure rates with the aim of reducing the wall friction between the auger and ceramic body as this creates a water film at the surface of the cooled auger.

In almost all cases, the cooling is done with a water–glycol mixture to prevent any freezing in the event of subzero temperatures, sometimes using heat transfer oil. The cooling temperatures usually range between minus 10 and plus 10 °C. Care must be taken that the difference in temperature over the cross section of the ceramic column must be kept as low as possible. It, therefore, makes sense to differentiate the cooling of the auger against the cooling of the barrel either by selecting different cooling temperatures for auger and barrel or by varying the flow of the coolant. No cooling but heating instead is applied in processing thermoplastic ceramic bodies, just to mention that at this point. In this connection, we do not want to go into the exact mechanism of cooling units which can be found in the relevant literature.

11.4.2 Selection of the Cooling Unit

The cooling machines employed for the re-cooling feature a cooling circuit that extracts heat from the circulation medium with the use of compressor technology, whereby cold water with the desired flow temperature is generated and is supplied to the consumer via pumps. The generated waste heat can be dissipated with the use of cooling water of higher temperature or with air. For selecting the suitable cooling unit, certain system decisions have to be made to suit the conditions at the site:

- A decision has to be made between water-based or air-based systems.
- With water-cooled systems, cooling water with say 25–30 °C is available for re-cooling of the cooling unit. With air-based systems, the energy is released into the atmosphere.
- Preference should be given to water-based systems provided that a cooling water piping is available at the site. In this case, the cooling unit can be positioned near the consumer, which will reduce the installation costs and pipeline losses.
- In the absence of cooling water, the choice must be made whether the cooling unit—compressor and capacitator—for an air-based unit will be positioned inside

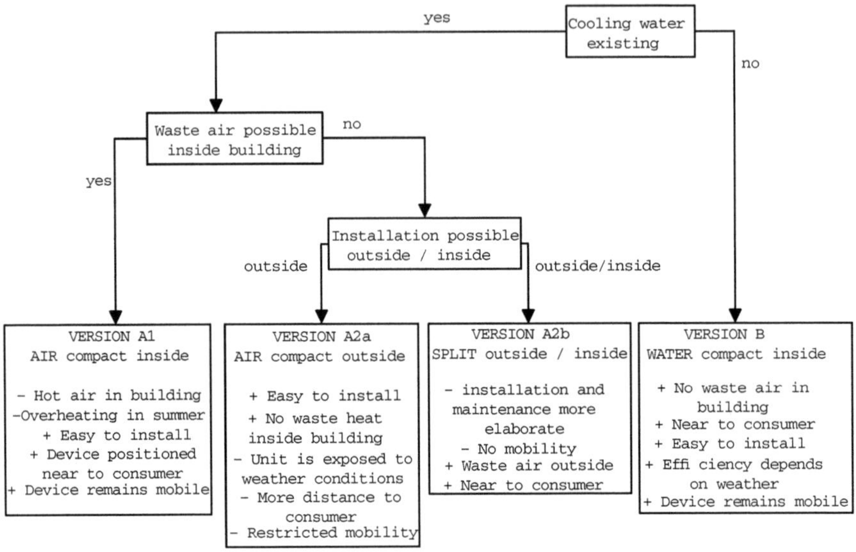

Fig. 11.1 How to choose the right cooling system

or outside the building. The disadvantage with indoor installation is the heating up of the room by the warm waste air.

- In the event of outdoor installation, weather protection may be required for the machine, plus insulated cables; moreover, a collection tray for the ethylene glycol is compulsory.
- There is also a so-called split solution with the waste heat-generating capacitator positioned outdoors whilst the compressor is located indoors. This entails increased installation costs for the refrigerant pipes between the components and restrictions as a result of the distance involved (Fig. 11.1).

11.4.3 The Major Operating Data for the Cooling System

The correct design of the cooling unit necessitates meticulous planning; the following information must be known:

- Throughput volume of the cooling water (m^3/h)
- Temperature of the cooling water (°C)
- Pump pressure (between 4 and 6 bar subject to conduit)
- Ambient temperature—max/min (°C)
- Water temperature in the locally supplied cooling water piping (°C).

11.4.4 *Also to Be Noted*

As a result of the cooling of the barrel and auger, and sometimes also of the extruder machine body, condensation water is inevitable. Suitable insulation of the components can not only prevent condensation water, but also enable the required cooling capacity to be substantially reduced.

11.5 Vacuum Pump and Vacuum Monitoring

11.5.1 *Definition of the Vacuum Pump*

If a space is free of matter we speak of "absolute vacuum", in so far as that is the physical definition. When extruding ceramic bodies, the aim must be to remove, as far as necessary, any trapped air from the ceramic body. Why? The reason is that during extrusion not only the ceramic body is compacted under pressure, but so is the air, from which the tension is abruptly released when exposed to the atmosphere after leaving the die. This is the reason for employing vacuum pumps, which are generally part of the ancillary equipment for an extrusion plant. Sometimes when extruding ceramic pellets and also in the extrusion of pottery, there is no need for de-airing.

The nominal displacement required for the vacuum pump (in m^3/h) does of course depend on the throughput rate of the extrusion plant, whereby it must be borne in mind that steam may be generated during the de-airing process. Some reserves must also be allowed for subject to the operation mode. The benefits of vacuum pumps with over-dimensioned displacement are that unexpected vacuum failures which may be caused by inadequate material feed can be compensated, and the disadvantage, on the other hand, is that air is permanently pulled through the vacuum chamber where systems are not airtight, which can generate dry agglomerates. If such vacuum failures occur frequently as a result of operating conditions, a more favourable economical solution would be the installation of the so-called low-pressure vessels in lieu of an over-dimensioned vacuum pump. Drying up of the ceramic body can also be caused by extended operational stoppages during which the vacuum pump has not been switched off.

The second criterion for the design of the vacuum pump is the so-called ultimate pressure (mbar). Extrusion of ceramic bodies does not require a high vacuum but rather a so-called rough vacuum ranging between 300 and 1 mbar, usually between 20 and 100 mbar. In Fig. 11.2, we have worked out a guideline to determine the size of the vacuum pump as a function of the output of the extrusion unit.

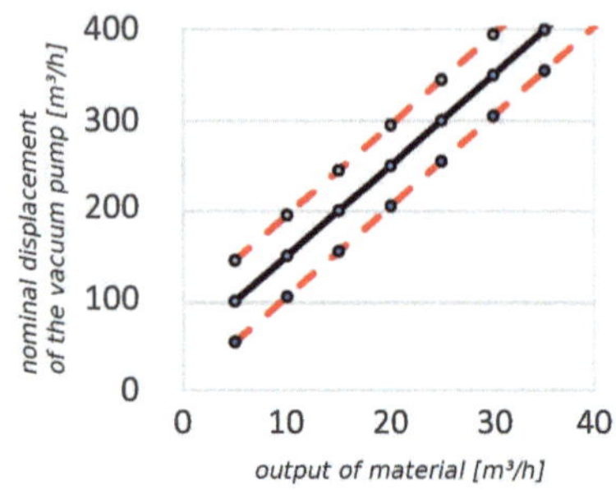

Fig. 11.2 Relation between the nominal displacement of the vacuum pump and the output of the extrusion unit

11.5.2 How to Measure the Level of Vacuum Correctly

In practise, mostly air-pressure-dependent barometric measuring instruments are used which show the pressure inside the vacuum chamber depending on the atmospheric pressure. Since these indicate different readings subject to the installation altitude, readings taken at the same pressure inside the vacuum chamber vary; in practise, this often causes confusion. Therefore, also in view of inevitable air-pressure fluctuations, it is recommended to use pressure-independent instruments for measuring the vacuum.

As an example: in Fig. 11.3, you can see the difference in the reading between barometric pressure-dependent reading and non-barometric pressure-independent reading. At a height of 1000 m and a temperature of 40 °C, the pressure is detected with the barometric measurement at about 190 mbar, whilst with the pressure-independent measurement, it shows about 75 mbar.

11.5.3 Types of Vacuum Pumps

Various types of vacuum pumps are employed subject to the throughput rate and required ultimate pressure. In ceramic extrusion, essentially two types have prevailed: rotary vane vacuum pumps and liquid ring vacuum pumps which can be operated either with water or oil. For brick factories with their robust operating conditions, the oil-operated liquid ring vacuum pumps are the preferred type; for applications where ultimate pressures of $\leq$100 mbar are needed, the rotary vane vacuum pumps are predominantly used (Fig. 11.4).

In the simplest case, ceramic bodies consist of a mixture of ceramic particles, water and air. However, the physical state of water changes as soon as the body enters the vacuum chamber. It is generally known that water turns into steam as the temperature rises and the pressure drops. Do you remember the saturated steam chart? For example: assuming the 1 l of water contained in the ceramic body has a temperature of 20 °C and the residual pressure in the vacuum chamber is 23 mbar; the result is that the water evaporates and 58 m^3 of saturated steam are generated from that

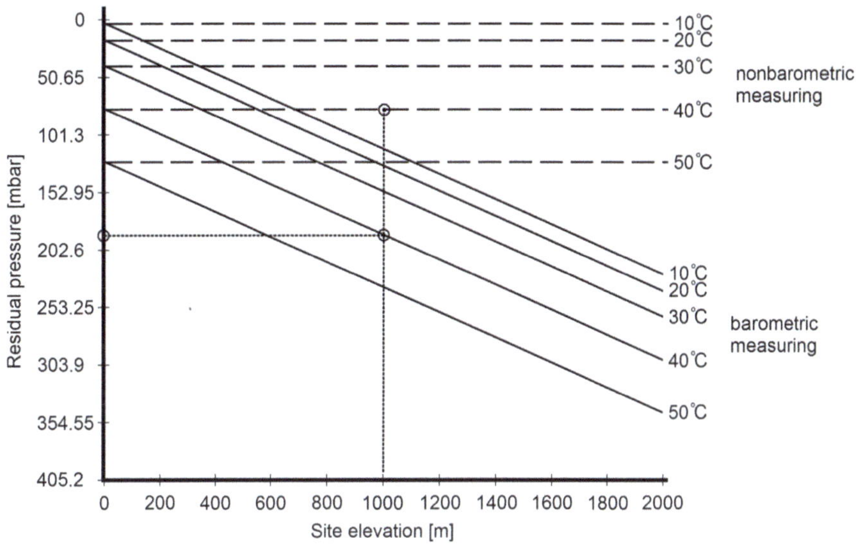

Fig. 11.3 Vacuum as a function of the site elevation and the temperature of the ceramic body

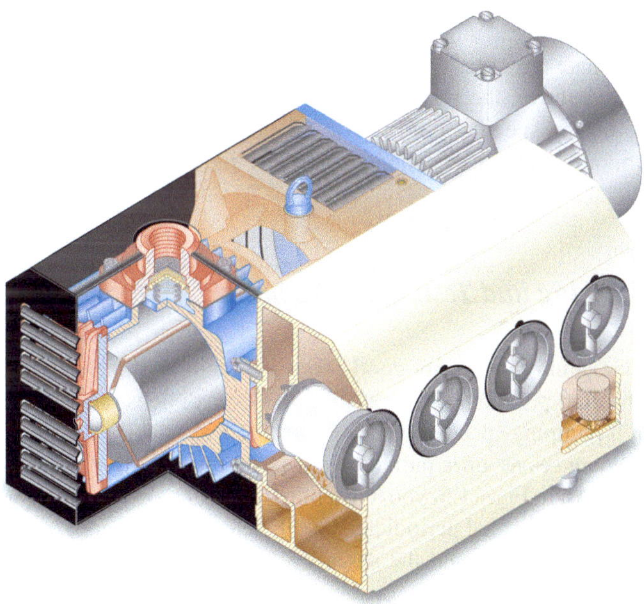

Fig. 11.4 Rotary vacuum pump (Gardner Denver GmbH)

1 l of water! If the water contained in the ceramic body had a temperature of 50 °C, as sometimes is the case with the so-called hot extrusion of bricks, the water would already evaporate at a residual pressure of 123 mbar. Hence, two consequences can be derived from this: too high a temperature of the ceramic body, at a high vacuum, cause a drying of the body in the vacuum chamber. The colder the body and the embedded water, the higher the vacuum can be for de-airing.

11.5.4 Regulation and Measurement of the Vacuum

There are special cases for which the de-airing process must be regulated and controlled. If the ceramic body is very moist, i.e. water and steam are de-aired, the vacuum pump must be additionally protected by liquid separators. All relevant details can be obtained and investigated on the respective websites of the vacuum pump suppliers; here is a summary of the key options:

- Manually or automatically controlled vacuum regulating valves are employed to lower the ultimate pressure to a preset value maintaining the same pumping speed. This prevents drying of the body during de-airing.
- Frequency-operated vacuum pumps are used to vary the pumping speed at a constant ultimate pressure.
- As mentioned above, the shown vacuum pressure is linked to pressure gauges depending on the specific atmospheric air pressure. Control and regulation of the vacuum only make sense when air-pressure-independent pressure gauges are used.

11.5.5 Checking and Testing the Vacuum

Even the best vacuum plant is not good enough if there exist leaks in the extrusion system which often pass unnoticed because the over-dimensioned vacuum pumps with their high pumping speed compensate for the leaks. This "compensation" has negative consequences as air is pulled permanently through the system resulting in dry agglomerations. In Fig. 11.5, you find a practise-oriented description of a vacuum test that should be performed periodically. Leak detection systems of different types are available in the market, and the so-called helium leak testing units have been found to achieve the best results.

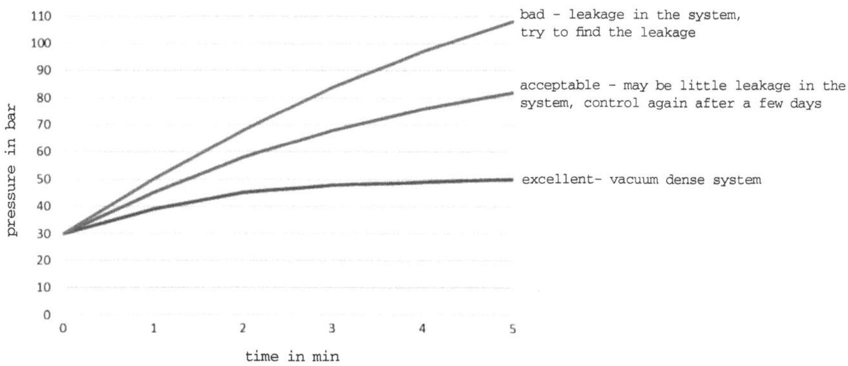

Fig. 11.5 How to detect leakage in the extrusion system

11.6 Hydraulic and Pneumatic Equipment

In the event of electric drives being replaced with hydraulic drives including the pump units, appropriate provisions must be made with respect to the cable runs, space requirement, operating materials, etc. The same applies to pneumatic plants. For both options—pneumatic or hydraulic—care must be taken that the lines from the generators to the consumers are not too long as this will cause line losses.

11.7 Screen and Die Changers

Hydraulic or manual systems to change the screen together with the strainer plate in front of the pressure head or to change the die are mainly applied to increase the efficiency of the extrusion plant by reducing the set-up time for changing the screen or the die. Whilst it takes at least 15 min to change these components, it takes less than a minute with a hydraulic screen changer, a big saving (Fig. 11.6).

11.8 Infrastructure

It happens quite frequently that the requisite infrastructural measures relating to the installation and operation of an extrusion plant are underestimated. The following summary shows the most important measures:

- Foundation and base frame
- Channels (ducts) for all types of lines
- Lifting gear for maintenance work such as change of augers, barrels, pressure heads and dies

Fig. 11.6 Modern hydraulic screen changer

- Water, pneumatic, hydraulic and electrical connections should all be within reach of the extrusion plant
- Measures for tempering and relative humidity of the environment
- Standard and special tools for maintenance and cleaning
- Vehicles such as forklifts, lift trucks and lifting platforms
- Industrial safety measures of all kinds.

11.9 Conclusions

- Do not forget or neglect the accessories in course of the implementation of an extrusion plant.
- Select the accessories with the same care as applied to the extrusion unit itself.
- Insist upon disclosure of the criteria relating to the dimensioning and design of the vacuum plant and the system for tempering the cooling circuit.
- Do not underestimate the requisite infrastructural measures to be initiated before the extrusion plant is installed.

Bibliography

1. W. Bender, D. Lutz, Special accessories for de-airing extrusion units. Keramische Zeitschrift **36**, 301–305 (1984)
2. ECT-newsletter 07/2010

3. F. Laenger, Evacuation in ceramic extrusion—dependencies and local situations, in *Extrusion in Ceramics*, ed. by F. Händle (Springer, Berlin, 2007), pp. 275–279
4. M. Redmann, Evacuation technology for ceramic extrusion, in *Extrusion in Ceramics*, ed. by F. Händle (Springer, Berlin, 2007), pp. 281–294

Chapter 12
Pressure Heads and Dies

Everything is going to be fine in the end. If it's not fine it's not the end.
—Oscar Wilde

Abstract How should the pressure head/die combination of the extruder be designed to prevent the internal stresses generated during the shaping of the ceramic body? Three prototypical examples—a highly perforated block, a stiff-extruded face brick and a ceramic honeycomb—are taken to demonstrate the difference in the design of pressure head/die combinations. Reference is made to conventional simulation methods as already adopted for the extrusion of plastics and which will be relevant in the future for ceramic extrusion.

Keywords Pressure head · Die · Brick extrusion · Stiff extrusion · Honeycomb extrusion · Flow simulation

12.1 What Matters

If you have read about the formation of laminations in Chap. 9, you are already aware of the essential aspects relative to the design of the pressure head and die, namely the ability to extrude ceramic profiles of the most diverse geometry with the least possible internal stresses. Such stresses manifest themselves later as laminations in the dried or fired product. The majority of these laminations are actually generated in the pressure head and die during the reshaping process of the ceramic material. However, there are cases where laminations are generated by other factors which are described in Chap. 9, the main reason for the generation of laminations is the shaping deformation of the body, and this process takes place in the pressure head/die combination. Conclusion: the quality of the very end of the extrusion unit has a decisive influence on the quality of the extrudate.

This leads to the question of how the manufacturers of pressure head and die combinations manage to generate an extrudate that flows in a virtually uniform

F. Händle, *The Art of Ceramic Extrusion*, https://doi.org/10.1007/978-3-031-63471-0_12

manner over the entire cross section with a minimum of internal stresses. In his contribution "Dies, Pressure Heads, Strainer Plates and More" (2009), Berger summarizes the key factors that are of essence for the design of such combinations.

Below are his explanations in a slightly modified form:

- The rheological properties of the ceramic body must be kept in mind for designing and building extrusion tools. These properties constitute the sum of numerous determining parameters—factors to which the raw material is exposed from preparation through extrusion.
- The influence of the size of the extruder and the design of the auger and extruder barrel, position and geometry of the end-auger, clearance between auger and barrel and the evacuation.
- The subsequent treatment of the green extrudate, i.e. handling, drying, firing.
- The requirements for the final product like ultimate shape, requirements regarding precision, surface quality, standards and approvals.
- The characteristics of the materials used for the structural components of the tool and the required output numbers for the different products.

So far for the findings of Berger, who had a lot of experience in designing pressure head/die combinations. Now let us take a look at some prototypical examples to demonstrate what is to be understood by a pressure head/die combination.

12.2 Typical Pressure Head/Die Combinations

12.2.1 Example 1: Pressure Head/Die Combination to Produce Large Brick Blocks

A typical pressure head/die combination for the extrusion of large clay brick blocks as developed in Europe is made up of dozens of components. Essentially, this is the pressure head itself, which is usually of a tapered sometimes enlarged shape; the bracket for the core holders integrated in the pressure head to which the cores are fitted, which produces the desired hole pattern, as well as various frames which define the external contour of the brick block (Fig. 12.1).

This means that, on the one hand, such pressure head/die combinations are always customized to suit the material and geometry of the extrudate. On the other hand, the dimensional accuracy and quality of the extrudate depend on the assumption that the aforementioned prerequisites are met. Should these preconditions change—such as the shrinkage of the body—any losses in quality and dimensional accuracy have to be accepted, or the pressure head/die combination must be adapted to suit the new prerequisites. In practise, such adjustment processes always happen in both directions: by generating a body of a constant, respectively, optimized consistency, and/or by adapting the pressure head/die combination to the prevailing conditions, be they influenced by the properties of the ceramic body, by the changes of the geometry

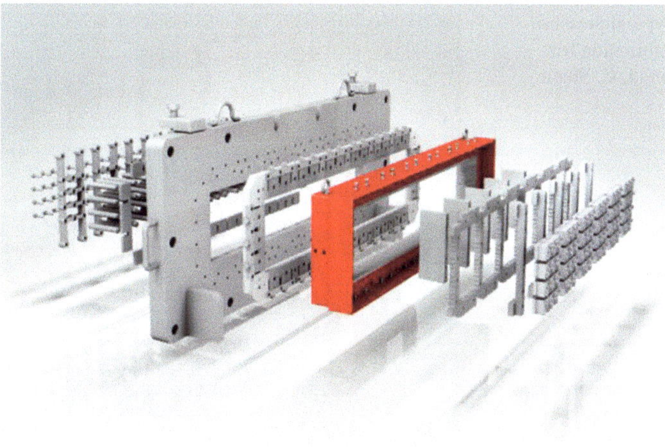

Fig. 12.1 Illustration of a typical pressure head/die combination for brick blocks

of the extrudate or by the application of new materials and coating methods used to reduce the wall friction and improve the wear characteristics.

12.2.2 Example 2: Pressure Head/Die Combination to Produce Stiff-Extruded Face Bricks

In the brick industry, so-called stiff extrusion is a process that allows the green extruded bricks to be set directly onto the dryer/kiln cars in several layers. This is only possible by adopting a relatively stiff extrusion method—or expressed in other words: if the deformation pressure reading is measured with the penetrometer—see Chap. 6—is relatively high. Stiff extrusion is a very interesting process assuming the material in the pit is not too wet and the perforation of the bricks is not too high—see Fig. 12.2.

A special feature of pressure head/die combinations as employed for stiff extrusion—as opposed to pressure head/die combinations for large clay brick blocks—is the lubrication system for the pressure head to reduce the wall friction between the clay body and the walls of the pressure head and die. The pressure rates usually experienced in extruding large clay brick blocks lie in the region between 20 and 30 bar, whereas the pressure may well be above 50 bar in case of stiff extrusion. The dosing of the lubrication liquid depends on the geometry of the brick, the output and the pressure in the pressure head/die combination and has to be done very carefully.

Fig. 12.2 Typical pressure head/die combination for stiff extrusion (J. C. Steele Inc.)

12.2.3 Example 3: Typical Pressure Head/Die Combination to Produce Honeycombs

In 1971, Corning Glass Works was granted Patent No. US3790654 A under the heading of "Extrusion Method for Forming Thin-walled Honeycomb Structures". Rodney Bagley was named as the inventor with Irwin Lachmann and Ronald Lewis involved in the team. For the first time, it was possible to extrude very thin-walled and complex honeycomb profiles from ceramic materials. In terms of design, this type of die is fundamentally different from the usual constructions as there are no core rods or any holders for the core rods and also no cores in the customary sense (Fig. 12.3).

To date, all honeycombs of differing sizes and geometries are produced according to this principle. Tests to extrude also other ceramic products such as highly perforated bricks with such dies have so far failed. A pressure head/die combination for honeycombs consists of:

- Firstly, a pressure head in one or two parts is usually chilled and provided with a pressure sensor ahead of the die plate, plus an adaptor ring fitted between the barrel and pressure head with a second pressure sensor to measure the pressure ahead of the integrated strainer plate.
- Secondly, a strainer plate is integrated into the pressure head with a screen to eliminate any agglomerations which would immediately cause unacceptable imperfections and rejects.
- Thirdly, the monolithic die plate, with inlet openings on the back as well as the "mini-cores" on the front, plus the frame to define the external contour.

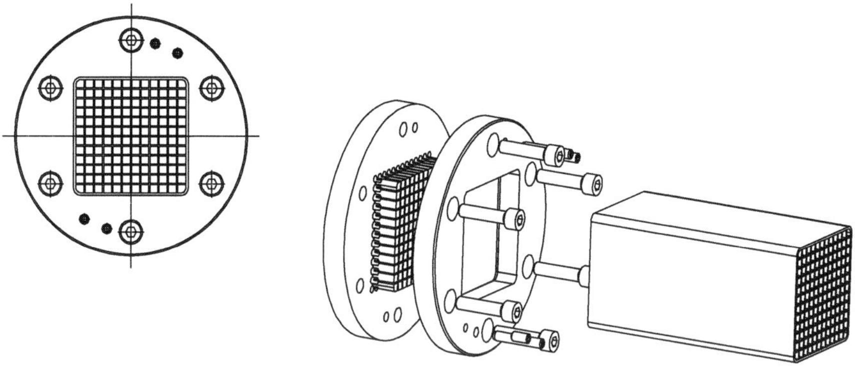

Fig. 12.3 Pressure head/die combination for honeycombs (ZM BRAUN GmbH)

Yet even with this assembly, which appears relatively simple, the devil is in the detail and it is rarely possible to produce perfect honeycombs at the first attempt without subsequent “fine-tuning”. Producers of honeycombs have invested substantial know-how in the design of the die plate—relationship of inner slot width to core sizes and feed holes—moreover in the internal contour and infeed angle of the pressure head as well as the temperature control of pressure head and die. Due to the fact that such dies are generally wire-cut EDM respectively sink-eroded, a decisive role is played by the precision achieved in machining and the surface roughness during this process. If, for example, we take a die with 500 cpsi and a diameter of 12 inch, there are about 5600 inlet openings and “mini-cores” which need to be machined.

12.3 A Look Over the Fence

Let us come back to the question of how the manufacturers of pressure head/die combinations manage to generate a uniform flow of the extrudate over the entire cross section with a minimum of internal stresses. What theories, methods, instruments and experience are available to us these days to assist in the optimum design of pressure head/die combinations for ceramic extrudates? It is worth taking a look over the fence. Extrusion as a continuous shaping method is not only restricted to ceramic bodies, but also employed in many other fields: for the extrusion of foodstuffs and animal feeds, for the extrusion of metallic materials, for pharmaceutical products and biomass, and above all for the extrusion of different types of plastics. From a rheological point of view, plastics are fundamentally different from ceramic compounds; they are predominantly extruded by the thermoplastic method and cannot be compared with ceramic bodies in terms of abrasiveness and corrosion. Yet likewise in the extrusion of plastics, it is important to minimize internal stresses in the extrudate. For the design of tools, nowadays CFD simulation methods are mainly used, such as the well-known multi-stage fluid dynamics solutions of ANSYS, which is not only

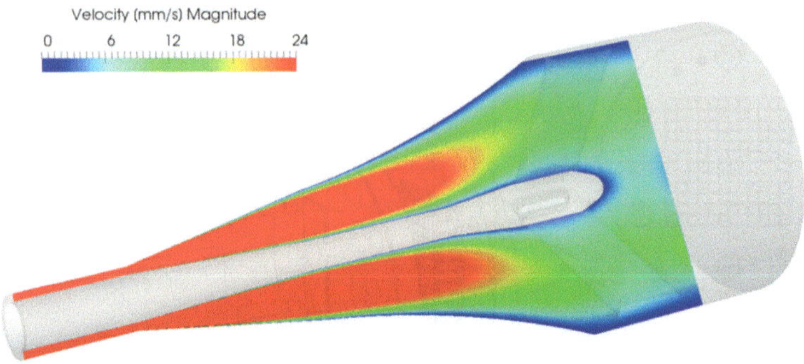

Fig. 12.4 Simulation of the velocity distribution of the ceramic body in a pressure head/die combination for tubes (IANUS GmbH)

used for the design of the tool, but also for strength optimization of the components, for registration of temperature distribution within the extrudate, hence universally applicable (Fig. 12.4).

Other special programmes focus on the design of augers or analyse the particle distribution. Attempts to apply for these programmes on a broad front for the simulation of the flow processes of ceramic materials or for the design of pressure head/die combinations have all failed up to now. This has not, however, excluded some manufacturers from successfully adapting these programmes for their own purposes. If one tries to understand why this is the case in spite of the astonishing performance of modern simulation methods, there appear to me to be several reasons. One is that there is a much higher demand for such applications in extruding plastics; another is that the temperature control with thermoplastic extrusion offers a correcting variable for precise control of the rheological properties, and finally, the wall friction is not a dominating factor in simulating the flow processes of plastics, this, however, is the case with ceramic bodies.

Up to now tools used for ceramic extrusion are primarily designed on the basis of many years of experience and empirical findings. This comes together with fine-tuning, trial and error and the hope that one of these days it will be possible to use efficient and practise-oriented simulation methods also for the extrusion of ceramics.

12.4 Conclusions

- The more accurately the rheological parameters of the ceramic body and the general framework conditions are known, the greater the chance of designing optimized pressure head/die combinations.
- Variations in the rheological properties of the ceramic bodies to be extruded or in the framework conditions lead to suboptimal results in extrusion which can only be remedied by adequate adjustments to the pressure head/die combinations.

- Modern simulation methods will eventually prevail also for the design of pressure head/die combinations for ceramic materials. It is, however, a precondition that the relevant rheological parameters of the ceramic bodies be available.

Bibliography

1. J. Wight, Honeycombs, in *Cellular Ceramics*, ed. by M. Scheffler, P. Colombo (Wiley-VCH, Weinheim, 2005), pp. 56–99
2. ECT Newsletter 05/2009
3. P. Steele, Soft and stiff extrusion compared, in *Brick and Tilemaking*, ed. by W. Bender, F. Händle (Bauverlag, Wiesbaden, 1982), pp. 236–240
4. H. Berger, Dies, pressure heads, strainer plates and more, in *Extrusion in Ceramics*, ed. by F. Händle (Springer, Berlin, 2009), pp. 221–230

Chapter 13
Extrusion Problems and How to Solve Them

All Life is problem solving.
—Karl Popper

Abstract The main problems of ceramic extrusion are listed together with the most probable causes for the problems and the most useful remedies.

Keywords Extrusion · Problems · Checklist · Problem-solving

13.1 A Bit of Help for Self-Help

Going back to Chap. 5, you will find a questionnaire which we have devised and which will help to give you a fairly accurate idea as to the type of extrusion plant you will need. This has been based on the assumption that you know the type of raw material to be processed, the products to be made, in which quantities and the working times including the major boundary conditions. There is another checklist in Chap. 18 which should also aid you in verifying prior to your final decision which of the short-listed extruders will best meet your particular requirements. Here in this chapter, it is assumed that you are already involved with extrusion but are now and again confronted with certain problems. The following checklist should assist you in solving extrusion-typical problems by way of a systematic approach. This list had to be restricted to the most significant cases which occur in practice. In this connection, I would like to draw your attention to the section "Problems" in the well-known book of James S. Reed, "Ceramic Processing"; in my 1995 edition, page 473 et seq. Any specific problems which are related to your existing extrusion plant should be addressed directly to the manufacturer of the extrusion equipment.

But any kind of problem solution necessitates the existence of reliable information in respect of all relevant operating parameters such as extrusion pressure rates, current consumption, degree of vacuum, column speed and torques. Regrettably, this is in many cases not available!

Often enough we experience that a de-airing extrusion unit has functioned absolutely perfectly for months after being delivered, and all of a sudden, more and more

F. Händle, *The Art of Ceramic Extrusion*, https://doi.org/10.1007/978-3-031-63471-0_13

problems accumulate for no apparent reason. We wonder why? Only in a few cases, one receives satisfactory answers and, in the rarest instances, is informed of changes that have been made since successful start-up of the plant, for example, modifications to the body, the condition of the wearing parts, variations in the material moisture content. You may find that the suggested measures such as checking of the wear rate of the auger and barrel liners repeat over and over again. In reality, very diverse problems originate from just a few causes: rough auger surface, variations in the moisture content of the prepared material, leaks in the extrusion system, etc.

The following checklist assumes that the boundary conditions ahead and after extrusion are right, which means that the dosing and preparation system upstream of the extruder provides the required performance and also that the handling and, in particular, the drying process downstream of the extruder are attuned accordingly (Table 13.1).

Table 13.1 How to solve the problems in extrusion—a checklist

Problem	Causes	Remedy
Insufficient output	Screw worn	Refurbish or replace screw
	Cylinder liners worn	Replace or refurbish cylinder liners
	Surface of screws is too rough	Polish the surface of the screw and/or renew the hard-faced layer
	Unadapted screw revolution—too fast or too slow	Check whether it is possible to adjust revolution of the auger according to the required output. Try to extrude "hungry" Check if the column speed reacts linearly with the revolution of the screw
	Perforated plate and mesh in the pressure head create a too high pressure drop	Check perforated plate, replace as needed/if possible, and check for blockage of perforated plate or mesh
	Inadequate geometry of the cylinder liners	Check if the geometry is suitable to stop rotation of the body
	Inadequate geometry of the screw	Check if the geometry is suitable regarding output and pressure generation
	Material moisture level—too high or too low	Control the moisture content

(continued)

Table 13.1 (continued)

Problem	Causes	Remedy
	Material sticks on the auger in the feeding zone	Polish auger, change geometry of the auger and control interaction between auger and feeding roller
Power consumption and temperature of the motor—too high	Surface of screws is too rough	Polish the screw
	Inadequate geometry of the screw	Check if the geometry is suitable regarding output and pressure generation
	Motor and frequency converter are defect or not aligned	Control function of motor and frequency converter including alignment
	Material—too stiff	Check the moisture content
	Foreign objects in the screw	Open extruder and clean
	Oil with incorrect viscosity or contaminated oil	Check oil type and contamination, and replace as needed
	Ambient temperature—too high	Instal oil cooler
	Bearings defect	Check the bearings and replace
Oil loss	Defective seals at shaft bearings	Replace shaft seals and races as needed
Vacuum—too low	Seal between cylinder and extruder case damaged	Check seals, and replace if necessary
	Seals between pre-extruder/ pug sealer and extruder damaged	Check seals, and replace if necessary
	Holes in perforated sealing plate before vacuum chamber—too large	Use sealing plate with smaller holes
	Too little or wrong oil in the vacuum pump	Refer to the operating manual of the vacuum pump
	Vacuum pump with insufficient pumping speed and/or ultimate pressure	Check vacuum pump
	Unknown vacuum loss	Organize leakage control and vacuum test (see Chap. 11)
Backflow in the vacuum chamber	Output of pug sealer—too high	Adjust output of pug sealer to match the extruder
	Fluctuating moisture of body	Check the moisture content of the body and adjust
	Length/diameter ratio (L/D) of the screw—too short	Check the L/D ratio and adjust if necessary

(continued)

Table 13.1 (continued)

Problem	Causes	Remedy
	Foreign objects in the die or perforated plate	Check die and remove foreign objects if necessary
	Free cross section and/or pressure drop of the die—too high	Control geometry of screw
Dried material in the vacuum chamber	Leakage of the vacuum chamber	Control vacuum chamber with leakage inspection device
	Temperature in vacuum chamber—too high (water evaporation)	Check temperature, and reduce temperature or vacuum as necessary
	Vacuum too high or too low	Use vacuum regulating valve
Temperature of the body—too high	Cylinder liners worn, and incoming temperature of the body—too high	Replace cylinder liners, and control incoming temperature of the body
	Screw surface—too rough or screw worn	Polish screw or replace screw
	Revolution of screw—too high	Adjust revolution
	Ambient temperature or incoming body temperature—too high	Instal adequate cooling of the screw and cylinder, and adjust incoming temperature of the body
Pulsing column	Screw worn	Replace or polish screw
	L/D ratio—too short	Check the L/D ratio and adjust if necessary
	Fluctuating moisture of the body	Check the moisture content of the body and adjust
	Revolution of screw—too high or too low	Adjust revolution of screw
	Insufficient output of pug sealer	Increase speed output of pug sealer
Laminations in the extrudate (see Chap. 9)	Rotation of the screw is seen in the column	Check geometry of the top auger, and enlarge distance between top auger and die (equalizing zone)
	Middle of the column advances faster than sides	Check geometry of the pressure head, enlarge distance between top auger and die (equalizing zone), and control angles in the pressure head and die

(continued)

Table 13.1 (continued)

Problem	Causes	Remedy
	Cuts on surface or rough extrudate after drying	Control geometry of pressure head and die, control grain size distribution and control drying conditions
Distortion of the extrudate	Die not aligned with the pressure head	Make sure that pressure head and die are aligned and centred
	One-sided roughness of the pressure head and/or die	Polish pressure head and die all over
Deviation in the required geometry of the extrudate	Shrinkage of the material was not considered in the design of the die, or shrinkage has changed	Control the shrinkage of the body or of the different bodies
	Extrudate expands after die because of the high elasticity of the body	Detect if and how much the body is elastic
	Revolution of the auger—too slow or too fast	Adapt revolution of auger
	Undulating surface of the extrudate	Provide equilibration zone after die, calendering after extrusion

13.2 Conclusions

- Try to analyse and understand your problem before acting.
- Make sure the processes of extrusion including the function of the vacuum unit are understood.
- Carefully study all the relevant manuals.
- Try to separate the influencing factors; i.e. do not fumble with several factors simultaneously.
- Do not fix it if it is not broken.
- Mistrust your glorious experience.
- Study the relevant literature regarding "problem-solving" and "decision-making".

Bibliography

1. R. Ackoff, *The Art of Problem Solving* (Wiley, New York, 1978)
2. J.S. Reed, *Principles of Ceramic Processing* (John Wiley, New York, 1995)
3. S. Shingo, *Quick Changeover for Operators: The SMED System* (Productivity Press, New York, 1996)

Chapter 14
Pelletizing by Extrusion

There is no smallest among the small but always something still smaller.
—Anaxagoras

Abstract Pelletizing is well known within ceramics in the area of material preparation as well as in the shaping of different pellet geometries. The following chapter offers an overview of the various pelletizing methods for ceramic bodies and describes the relevant basics and elements of pelletizing.

Keywords Pelletizing · Agglomeration · Trilobes · Matrix plate · Pelletizing plant · Radial extrusion of pellets

14.1 What Do We Understand by Pelletizing?

A pellet is a small compacted cylindrical, spherical, or possibly star-shaped body. The term pelletizing refers to the manufacturing process for the production of such bodies. The basic masses used are extremely different from each other and range from biomass, to foodstuffs, right through to ceramic materials, the latter being of interest to us here (Fig. 14.1).

In a broader process-engineering context, pelletizing is in fact a type of agglomeration. One distinguishes between a build-up granulation process with the aid of pelletizing tables or pelletizing drums and press pelletizing, e.g. by briquetting using briquetting rolls, the so-called matrix presses, tabletting presses, and then, there is the pelletizing method by extrusion (Fig. 14.2).

The fundamental difference between build-up pelletizing and the press pelletizing process is the higher compaction achieved by press pelletizing. The basic difference between briquetting, tabletting and pelletizing process using matrix presses on the one hand and the extrusion method on the other is the higher degree of plasticity of the bodies to be extruded. So far for a preliminary broad outline on the subject of pelletizing.

F. Händle, *The Art of Ceramic Extrusion*, https://doi.org/10.1007/978-3-031-63471-0_14

Fig. 14.1 Pellets for eating

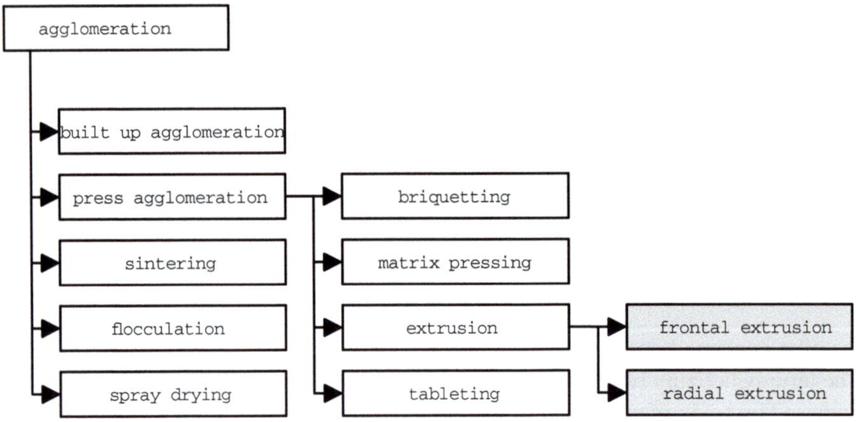

Fig. 14.2 Pelletizing by extrusion as one process of agglomeration

14.2 Pelletizing by Extrusion

Let us now deal with our main topic of pelletizing by extrusion. Ceramic masses of different consistencies are used to extrude pellets in the form of bulk catalysts, which for instance are used in huge quantities in the petrochemical industry. Moreover, they are used as an additive for building materials, for the storage of liquids, and also as filling material for the fracking process—the so-called proppants. Ceramic pellets are also applied as a primary product in the production of abrasives and superabsorbers (Fig. 14.3).

Bulk catalysts in particular will in the future be of great significance for environmental purposes. Companies like Haldor Topsoe, BASF, CRI, Umicore and Johnson Matthey produce up to 150 different types of bulk catalysts of various geometries,

Fig. 14.3 Extrusion of plastic rods

using different materials and for the most diverse applications. And many of these catalysts are produced by the extrusion process.

14.3 Plasticity as a Prerequisite for Pelletizing

The key requirement for pelletizing by extrusion is the existence of a minimum plasticity, which is defined by the material itself, the water content, the number and type of binders and additives. In the absence of such minimum plasticity, as might be the case in pelletizing of straw, wood or similar biomasses, other pelletizing methods must be employed. In this connection, reference should be made to the information provided in Chap. 6 on the subject of plasticity. If there is minimum plasticity, pelletizing by extrusion offers many benefits due to the fact that the cross section geometry can, within limits, be freely designed. As an example, the so-called trilobes or small tubes are often employed to enlarge catalytically active surfaces which cannot, or only to limited extent, be generated using other pelletizing methods (Fig. 14.4).

14.4 Pellet Geometry

As a rule, the lengths of the pellets are created stochastically by breakage caused by gravity when exiting the dies also called the matrix plate. If an accurate cut is required, then rotary or linear acting cutters are employed. In this case, the pellet length can be infinitely adjusted and varied by the number of cutting actions. In case of a predetermined L/D ratio, specially designed cutters must be used, or alternatively,

Fig. 14.4 Matrices for trilobes

the dried and sintered pellets need to be sorted and selected for length through the screening process (Fig. 14.5).

Once extruded, the pellets drop onto a belt to be transferred to the dryer, for example, a fluidized-bed dryer, and finally to either continuous or intermittent kilns for sintering (Fig. 14.6).

Fig. 14.5 Special pellet cutter (ROCHET SA)

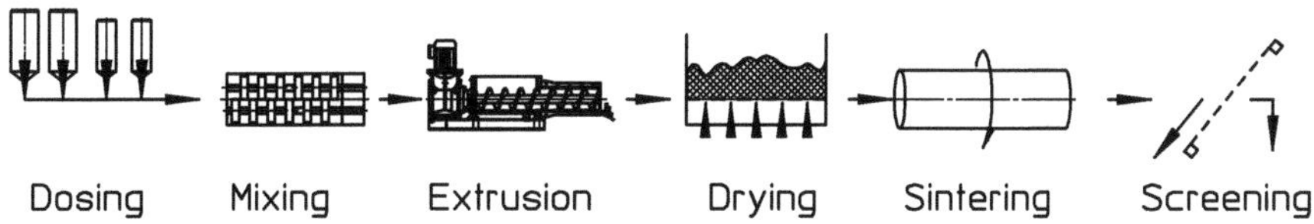

Fig. 14.6 Scheme for a pelletizing plant

14.5 Essential Features of a Pellet Extruder

What are the typical characteristics of a pellet extruder and which aspects should be taken into account? It makes sense to define a pellet extruder according to the following characteristics:

- Single- or twin-shaft extruder
- Maximum extrusion pressure
- Throughput rate
- With/without de-airing
- With/without cutter
- With/without matrix plate changer
- Number of dies and configuration in the matrix plate.

Apart from the size of the extruder as defined by the auger and barrel geometry, one can differentiate between extruders designed for low, medium and high pressure rates. This definition is arbitrary, but as a broad consensus, one could say that pressure rates—see Chap. 2. To suit the specific requirements for pelletizing, the extruders need to be equipped with an appropriate drive and suitably designed auger and barrel components for the required pressure rates. In this context, please refer to Chap. 4 (Fig. 14.7).

14.6 Dies and Matrix Plate

Bearing in mind that sometimes pellets feature extremely fine profiles with diameters of about 1 mm, great importance must be attached to the design of the so-called dies or matrices. These must feature a long life, low friction between the die and the ceramic body, and adequate stability to withstand high pressure rates. The design and choice of material of the dies is strictly protected know-how of the producers of pellets. Dies manufactured using plastic such as Teflon, also ceramic or steel materials and hard metal are to be found. However, not only the dies themselves but also the matrix plates which carry the matrices must absorb high pressures and not be allowed to deform. Modern methods of calculation using the assumed pressure rate values as a basis offer sufficient accuracy to achieve this Fig. 14.8.

Fig. 14.7 Extrusion of pellets with large—10 mm (*left photograph*)—and small (*right photograph*)—1.2 mm—diameters (J.C.Steele Inc., ECT GmbH)

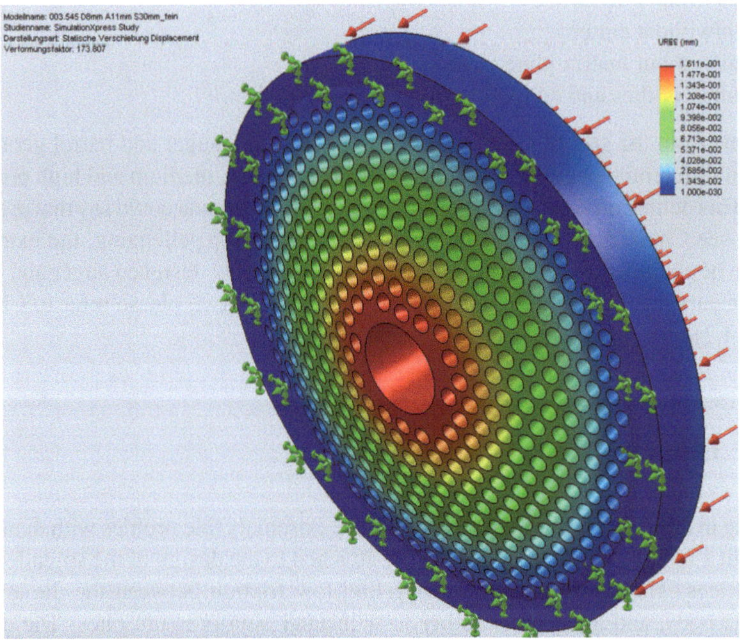

Fig. 14.8 FE calculation of a matrix plate

Fig. 14.9 Hydraulic matrix plate changer

14.7 Matrix Plate Changer

For high capacities, matrix plate changers are employed in order to avoid production disturbances caused by die cleaning. It has proven to be good practice to stop the extruder for a short moment and run the auger backwards by one to two turns for the purpose of reducing the pressure from the matrix plate, so that this can be shifted. If the second preinstalled matrix plate is provided, this operation takes less than half a minute and can be controlled automatically. In particular, when processing masses with a high percentage of fibres, or also in the event of very sticky masses which can quickly cause blocking of the dies, such screen changers considerably reduce the unproductive downtimes (Fig. 14.9).

14.8 Other Pelletizing Methods

So much for a brief summary on the subject of pelletizing by extrusion in the ceramic industry. To finish here are just a few additional notes. In practice, one can find units with frontal extrusion through a matrix plate arranged at right angles to the auger, but there are also extruders on which the pellets are discharged radially throughout the circumference or part of the circumference of a barrel. Finally, in the widest possible sense, circular screen feeders or perforated rolls as used in material preparation for the structural ceramics industry may be counted amongst the machines for press pelletizing. However, these should be considered as bespoke solutions which

Fig. 14.10 Radial extrusion of pellets (Händle GmbH)

only make sense for specific cases. Piston extruders can of course also be used for pelletizing, especially for small outputs and extreme pressure rates (Fig. 14.10).

14.9 Conclusions

- Define exactly all geometries of the pellets and the output rates for each type.
- Define the rheological characteristics, especially the plasticity—see Chap. 6—by a representative test in the laboratory.
- Try to estimate the pressure rates for the different pellet geometries to calculate the matrix plate.
- Select the right material for the matrices regarding the abrasiveness of the material and the performance of the extrusion plant.
- Select the right extruder.

Bibliography

1. S. Döring, *Pellets als Energieträger* (Springer, Heidelberg, 2010)
2. I. Kurunov, A. Bizhanov, *Stiff Extrusion Briquetting in Metallurgy* (Springer, Heidelberg, 2018)
3. W. Pietsch, *Agglomeration in Industry* (Wiley, New York, 2004)
4. H.F. Rase, *Handbook of Commercial Catalysts* (CRC Press, Boca Raton, 2000)
5. J. Regalbuto (ed.), *Catalyst Preparation* (CRC Press, Boca Raton, 2006)
6. H.B. Ries, Technologische Lösungen zur Granulierung und Pelletierung. In: Keramische Zeitschrift, pp. 208–210, 279–283
7. H. Schubert, *Aufbereitung fester mineralischer Rohstoffe* (VEB deutscher Verlag für Grundstoffindustrie, Leipzig, 1989)

Chapter 15
Wear, Tear and No End

Where is no friction there is no wear.
—Experienced Anonymous

Abstract To avoid wear is one of the main tasks in ceramic extrusion. The tribological basis of wear in its different appearances is described as well as the most relevant options for wear protection. In the complex wear mechanism, the roughness of the surface of the auger and the geometry of the liner are of paramount importance as well as the control of the gap between auger and liner.

Keywords Wear and tear · Corrosive wear · Abrasive wear · PTA welding · HVOF coating · Tribological system

15.1 Wear—A Real Challenge

The subject of wear and tear with its connected economic and technical aspects can rightly fill whole libraries. The number of industrial companies and research organizations who are engaged in the development of new materials, processes and coating methods is legion. Within our context, wear always results in higher operating costs, loss of productivity and quality deficiencies due to

- shorter service life of the parts.
- increased maintenance costs.
- unwanted contamination of the body.
- downtimes caused by component failure.
- changes in operating conditions, etc.

The reduction of wear and tear is consequently considered one of the major technical and economic challenges, not only in industry. Can we imagine a world where we have no wear—just think of our intervertebral discs and hip bones? I fear this will remain a dream. Let me list a few rough boundaries before we focus on the wear encountered in ceramic extrusion.

F. Händle, *The Art of Ceramic Extrusion*, https://doi.org/10.1007/978-3-031-63471-0_15

15.2 Types of Wear

In principle, we differentiate between:

- adhesive wear.
- abrasive wear.
- surface fatigue.
- fretting wear.
- erosive wear.
- corrosion and oxidation wear.

As far as extrusion is concerned, we will restrict ourselves in this context on the treatment of "abrasive wear" and "corrosive wear". This does not mean that the other types of wear in extrusion should not be observed; however, they play a minor role. In practice, there is almost always an overlapping of different types of wear, which could lead to trade-off situations in particular in the case of simultaneous occurrence of abrasive and corrosive wear. That means it is necessary to make compromises in solving the wear problems.

In the case of abrasive wear, there is a progressive loss of material at the surface of the solid body caused by mechanical stress as a result of the contact and relative movements of a solid, liquid or gaseous antibody, as stipulated in the German standard DIN 50302.

This situation is illustrated in the tribological basic system in Fig. 15.1 Without knowing the specific characteristics of the two "wear partners", solid body and antibody, as well as the ancillary conditions such as the type of movement and stress, it is impossible to develop a sustainable solution for the reduction of wear; it always remains trial and error.

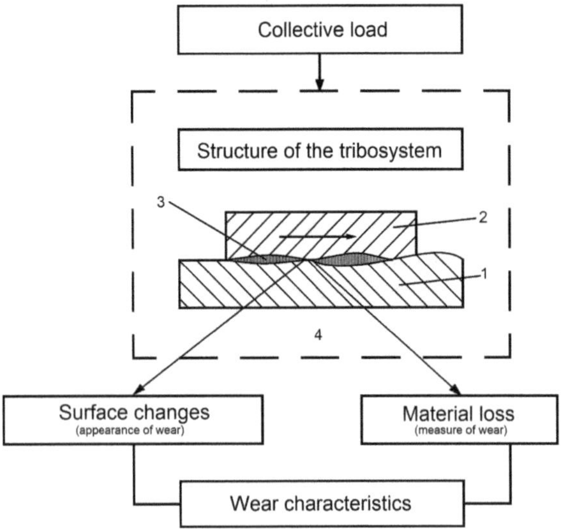

Fig. 15.1 Tribological system (1) primary body, (2) counter body, (3) interfacial medium, (4) surrounding medium

Corrosive wear is the result of chemical reaction between the corrosive ceramic body and the predominantly metallic parts coming into contact with that body. In view of the fact that ceramic bodies are frequently acidic or alkaline—pH levels >12 are not unusual—thechoice of suitable materials is of great significance. Apart from that, both corrosive and abrasive wear may cause contaminations in the body, which are frequently of an unacceptable nature.

15.3 Summary of the Most Important Wear Solutions

Table 15.1 gives the common options for wear reduction relating to the most important wearing part—the auger.

Next to the hardness of the material, respectively the coating, an attempt has been made to include in this list an indication for the roughness of the surfaces—see also Chap. 4. However, these indications must be interpreted with some reservation, bearing in mind that certain coatings are very smooth when new or in a polished state but will become rougher in operation, such being the case with PTA hard-facing or also with HVOF coatings using tungsten carbide. The hard carbides embedded in the binder matrix wear out less quickly than the softer matrix. This means that the softer nickel or cobalt matrix will wear out first but the particles remain producing a rougher surface (Fig. 15.2).

In general, the outside rim of the auger wears first, then the front flanks of the auger and finally the hub, but it depends not only on the position of the wings in the auger—feeding zone or pressure zone—but also the specific ceramic body, the pressure development over the length of the auger and the geometry of the auger. In some cases, the outside rim is still in perfect condition whilst the hub or the back flanks of the auger shows increased wear. That is exactly why different materials for wear protection are applied in different ways on the same auger, for example, small ceramic plates to protect the outside rims, PTA welding on the front flanks and HVOF coating on the back flanks and the hub.

With monolithic materials such as ceramics or hard metal, other factors besides the hardness, like the fracture toughness or break resistance, also play a decisive role. More information can be found in the relevant literature in particular concerning the so called Weibull factor which is of great importance in respect of ceramic components in calculating their failure probability.

15.4 Roughness of Auger Surfaces

A few years ago, we carried out an interesting test with the aim of evaluating in practice and under controlled conditions the influence of the auger roughness on the throughput rate when extruding ceramic materials. Obviously, everyone who has ever been involved with extrusion knows that worn augers reduce the throughput

Table 15.1 Options for wear protection of augers for extruders

Options for wear protection of augers	Outside rim	Front flank	Remaining surfaces (Back flank, hub)	Recommendation for …
A	WC/TC (PTA) 3 mm (700 HV)	WC/TC (PTA) 1.5 mm (700 HV)	WC/TC (HVOF) 0.2–0.3 mm (1700 HV)	… for highest abrasive exposure
B	WC/TC (PTA) 3 mm (700 HV)	Stellite 6 (PTA) 1.5 mm (400 HV)	Stellite 6 (PTA) 1.5 mm (400 HV)	… for abrasive and corrosive exposure
C	Stellite 6 (PTA) 3 mm 400 HV	Stellite 6 (HSP) 0.2–0.3 mm 400 HV	Stellite 6 (HVOF) 0.2–0.3 mm 400 HV	… for highest corrosive exposure
D	Stellite 6 (PTA) 3 mm 400 HV	Stellite 6 (PTA) 1.5 mm 400 HV	Stellite 6 (HVOF) 0.2–0.3 mm 400 HV	… for high abrasive exposure
E	WC/TC (PTA) 3 mm 700 HV	Hardchroming 0.5–3 mm 1000 HV	Hardchroming 0.5–3 mm 1000 HV	… for high abrasive exposure
F	Kolsterising	0.35 mm 1300 HV	0.35 mm 1300 HV	… only for small augers
G	PM composite	Depending on layer max. 800 HV	Depending on layer max. 800 HV	… for special applications—costs!!!
H	Hard metal	Monolithic max 1800 HV	Monolithic max 1800 HV	… for segmented augers/top augers
I	CVD	Layer up to 50μ max. 2800 HV	Layer up to 50μ max. 2800 HV	… for max. wear protection
J	PVD	Layer up to 50μ max. 2800 HV	Layer up to 50μ max. 2800 HV	… for max. wear protection

PTA Plasma-transferred arc welding, *HVOF* high-velocity-oxygen fuel, *WC/TC* tungsten carbide

capacity and reduce the pressure generating ability, but this was not the purpose of the exercise for us. The intention was rather to find out which throughput rates can be achieved under the same boundary conditions and in particular identical auger geometries, i.e. using augers which vary only with regard to their surface roughness of the auger flanks and the auger hubs.

The test was performed by extruding 200 cpsi honeycombs using SiC material through an extruder with a barrel diameter of 160 mm. The auger was welded on the outside rim with tungsten carbide (PTA), and the flanks and the hub were coated with tungsten carbide (HVOF). The honeycombs were extruded at a column speed of 3.5 m/min at an auger rotation of 12 rpm with a radial pressure of approximately 70 bar. The roughness of the auger surfaces showed a R_a value between 2 and 3.

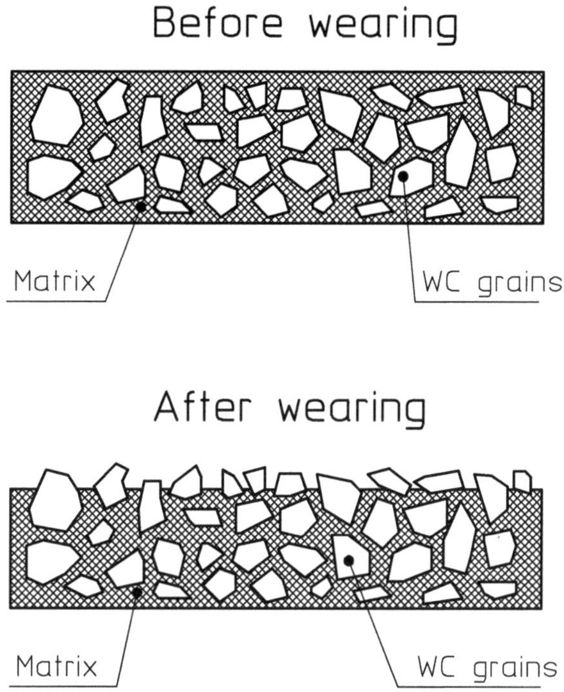

Fig. 15.2 Wearing of the matrix

In parallel, an identical auger was hard chrome-plated, although we knew that hard chromium plating for the extrusion of such an abrasive body offered no durable alternative. The roughness of the hard chrome-plated auger was between 0.3 and 0.4 R_a value. The result was that the same column speed of 3.5 m/min could be achieved with an auger rotation of only 4 rpm/min in comparison with the previous 12 rpm/min, i.e. so at one-third!

These findings were able to be verified by similar tests using different materials and profiles, subject to specific ceramic bodies, extrudates and boundary conditions in different relations. Rough auger surfaces can however also cause other problems. When using sticky adhesive materials, these may frequently stick to the auger hub mainly at the infeed zone. This can result in a complete blocking of the extrusion process due to the fact that the auger cannot receive any new material (Fig. 15.3).

Result: It is not only a question of generating hard and wear-resistant auger surfaces. It is the roughness of the surfaces and the long-term maintenance of a small gap between auger and barrel liner which in many cases are more important, also with regard to the product quality.

Fig. 15.3 Wear protection with ceramic plates

15.5 "Where There Is No Friction There Is No Wear"

This sounds quite obvious, but is it? If I remember it correctly, the first time I had a discussion about this was with Professor Günther Menning of University Chemnitz. Unfortunately, the devil lies in the detail, namely, in the boundary conditions of the tribological basic model and the quality of the basic body and the antibody. It is impossible here to go into any details. I would rather try to explain a practical consequence based on this rule. Whenever it is important to reduce contaminations caused by wear, one should also consider the use of soft and very smooth materials such as augers made of modern and high-strength plastics as against adopting extreme solutions, say augers of hard metal or ceramics. In such cases, it is a condition that the pressure rates are not too high in order to avoid any scratching of the hard ceramic particles into the auger surface. There are however other possibilities of reducing the "wall friction" between auger and ceramic mass. The two most important options are

- Addition of the so-called lubricants to the body such as glycerol, olein and graphite. In the majority of cases, the percentages of these additives to the body are <1%.

What can be an undesirable side effect is that the "internal friction" of the ceramic body changes, which may result in a varying flow behaviour of the body in the pressure head and die.

- Cooling of the auger, i.e. generation of a condensation film on the auger surface which also helps to reduce the wall friction between ceramic body and auger.

  Care must be taken that the temperature gradient over the cross section of the extrudate is kept as low as possible in order to avoid the so-called wet spots with increased viscosity in the centre of the extrudate—see also Chap. 9.

15.6 Function of the Barrel in Extrusion of Ceramic Materials

The rotating auger within the barrel can, in combination with a rough auger surface and worn barrel liners, cause the compact ceramic body to rotate with the auger in a circumferential direction. In order to avoid this, the wear liners in the barrel are provided with grooves of different geometries. Other options to avoid rotation of the ceramic body are to fit counter knives, the so-called pins, into the barrel protruding up to the hub. However, the disadvantage is that the level of efficiency in the extrusion is reduced and the so-called cutting textures could develop.

To sum up: the "ideal extruder" for ceramic bodies should be such that firstly the friction between auger and body must be kept as low as possible; secondly that the friction between the wear bushing and the body in circumferential direction should be as high as possible to minimize the rotation of the body and thirdly to keep the friction in the barrel liner in axial direction of extrusion as low as possible. Figure 15.4 shows the most significant options to stop the rotation of the mass in the cylinder: liners with pins also to increase the mixing effect (a), liners with rectangular grooves (b) and liners with "sawtooth" grooves (c).

Another essential criterion is to minimize the gap between auger and barrel liner. Worn rough augers or/and worn barrel liners causing an increased gap will reduce the

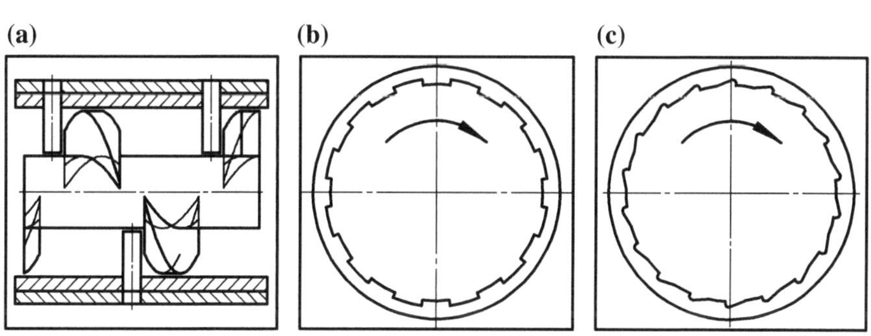

Fig. 15.4 Options to stop rotation

efficiency of the extruder. The narrower the gap—including the area of the grooves—plus the more edges there are to stop the rotation of the mass the higher is the chance to avoid "backflow", and the higher the efficiency of the extruder especially when applying high pressure rates and a high viscosity of the body. To this extent, barrel bushings with sawtooth profile grooves, either in spiral or longitudinal configuration, are the best solution in the majority of cases.

15.7 Conclusions

- Try to find augers as wear-resistant as necessary for the ceramic body and the pressure range and as smooth as possible.
- What kind of wear do you expect regarding your ceramic body: corrosive or abrasive wear, or both?
- Monitor the wear of the auger and barrel from time to time. Polish or regrind the worn auger and change the barrel liners if necessary; you can save a lot of money!

Bibliography

1. H. Czichos, K.H. Habig (eds.), *Tribologie-Handbuch* (Vieweg + Teubner, Wiesbaden, 2010)
2. G. Mennig, Tribological principles, in *Extrusion in Ceramics*, ed. by F. Handle (Springer, Berlin, 2007), pp. 313–320
3. I. Kleis, P. Kulu, *Solid Particle Erosion* (Springer, London, 2008)
4. E. Rabinowicz, *Friction and Wear of Materials* (Wiley InterScience, New York, 1995)
5. K.H. Zum Gahr, *Microstructure and Wear of Materials* (Elsevier, Amsterdam, 1987)

Chapter 16
The Ceramic Production Laboratory

The proof of the pudding is in the eating.

Abstract Why it is necessary to instal a production laboratory? About the different types of production laboratories. The science of sampling and the possible risks of insufficient control of the ceramic body before extrusion. A list of the basic equipment for the production laboratory.

Keywords Production laboratory · Sample testing · Basic equipment · Ceramic laboratory · Bending strength

16.1 What Are the Reasons for Having a Production Laboratory?

In principle, one should not have to justify the need for maintaining a ceramic laboratory as a production-accompanying support service. A laboratory not only serves as the responsible department for monitoring and ensuring quality control, but is also set-up to be used for developing new or modified ceramic bodies and products. And last but not least, it represents an institution within the company which strongly and permanently reinforces the importance of product quality and product innovation in the awareness of the staff and co-workers.

16.2 The Types of Production Laboratories

Basically, three models are in practical use depending on the size of operation and the type of products to be manufactured.

- In-house laboratory, properly staffed and equipped with suitable instruments, as well as having adequate ranking within the company hierarchy
- Outsourcing of the functions of the production laboratory to external laboratories

F. Händle, *The Art of Ceramic Extrusion*, https://doi.org/10.1007/978-3-031-63471-0_16

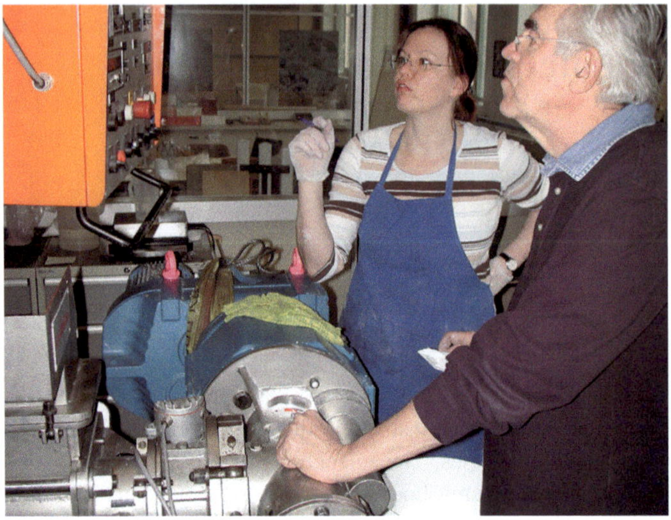

Fig. 16.1 Look in a ceramic laboratory (Händle GmbH)

- The merging of several related companies, for example, in the form of an association, to monitor quality control and the development of products.

Between these three basic models, there are other different and overlapping variations. For instance, it may make sense to entrust complex tests and R&D studies to external laboratories if the available in-house staff is not sufficiently qualified, or where the available devices are unsuitable, but with the basics being established in the in-house laboratory. Obviously, a laboratory to supervise the production of honeycombs is a completely different outfit than that of a laboratory for a brick factory. Yet the basic structure and the functions of both production laboratories are identical (Fig. 16.1).

16.3 Frequency of Sample Testing

16.3.1 When and How Often?

Unfortunately, there is no general answer to this question; that depends! It depends on which variations have to be expected with the raw material or body, the extent of these fluctuations, acceptable tolerances in terms of the product quality, etc. I know plants in which the laboratory takes certain samples on a weekly basis, I know plants where samples are tested every day, and I know plants in the technical ceramics industry where samples are tested and evaluated every hour. However, sampling is only the first step. As long as the detected readings are within the prescribed thresholds, it is just a matter of recording the readings, so that even after time lapses of months or years, the reasons for any possible deviations can be understood and reproduced. What however happens if the laboratory finds deviations which might result in quality defects?

16.3.2 The Worst Case

In practice, the pattern of decision-making can be very different. It can start from suppression of the problem with the motto "nothing will happen", through to panicky intervention by the management. Each type of reaction can be seen as reasonable, depending on the significance of the product quality and the consequences in the event of quality defects. Just imagine you have produced defect honeycombs to be fitted as three-way catalytic converters or diesel particulate filters (DPFs) in hundreds of thousands cars. Or perhaps a bit less dramatic: as a result of time pressure you have delivered roof tiles to cover a hundred roofs carrying a 30-year guarantee, but the roof tiles are found not to be frost-resistant. In order to be equipped for such emergency situations, it is worth examining the most important factors with the aid of a relevant decision scheme.

- Position of the production laboratory within the organization:
  Directly below top management with the obligation to report direct and not in a roundabout way, or settled just somewhere within the hierarchy jungle?
- Time lag:
  What is the time lag between detecting a possible defect and the time of intervention?
- Measures:
  Can the defects be corrected by accepting collateral damage if the intervention is carried out during operation, or are the defects irreversible and necessitate a production stop? (Fig. 16.2).

16.4 The Subject of Random Samples

In this context, we cannot avoid explaining the two terms used in random sampling, accurate random sample and representative random sample. Let me try to list only those essentials which are of importance to you in practice. In Wikipedia, the term random sample is defined as follows:

> In statistics, a simple random sample is a subset of individuals (a sample) chosen from a larger set (a population): each individual is chosen randomly and entirely by chance, such that each individual has the same probability of being chosen at any stage during the sampling process, and each subset of *k* individuals has the same probability of being chosen for the sample as any other subset of *k* individuals.

Fig. 16.2 Tests to produce large honeycombs (Keramikinstitut Meissen)

Further related literature can be found in Wikipedia under this keyword or in the relevant statistical literature—see references Chap. 15.

Let us assume that you are in a clay quarry where the raw material for your bricks is extracted. The driver of the bucket wheel excavator has been instructed to extract the clay over the entire cross section, which by no means is of a homogeneous nature, to be subsequently mixed in a large heap. At this stage, the first so-called sampling would take place. Would this be called a random sample? Perhaps one should ask the driver of the bucket wheel excavator!

Let us now assume that this happens similarly in two other quarries; i.e. three raw materials in three equal parts are prepared at the brick factory to form the basic raw material. The characteristics of the raw material, such as the mineral composition, the grain size distribution, the moisture content during extrusion, the proportion of inert and organic components, have been recorded in the laboratory. In this case, the task of the laboratory is to check, either at regular intervals or in the event of quality problems, whether the characteristics of the tested material are identical to those characteristics which are binding for the defined raw material compound.

Now and again, there may be complex conditions where you cannot avoid obtaining external help for determination of the random samples and the sampling, for instance in deciding on the number of random samples to be taken, the time gap between the sampling and the method of sampling, to ensure that all conditions for a representative random sampling are met. The German Standard DIN 51061: "Testing of ceramic raw and finished materials—Sampling of ceramic raw materials" is describing the procedure of sampling in detail. A great help! (Fig. 16.3).

Fig. 16.3 Sample splitter (RETSCH GmbH)

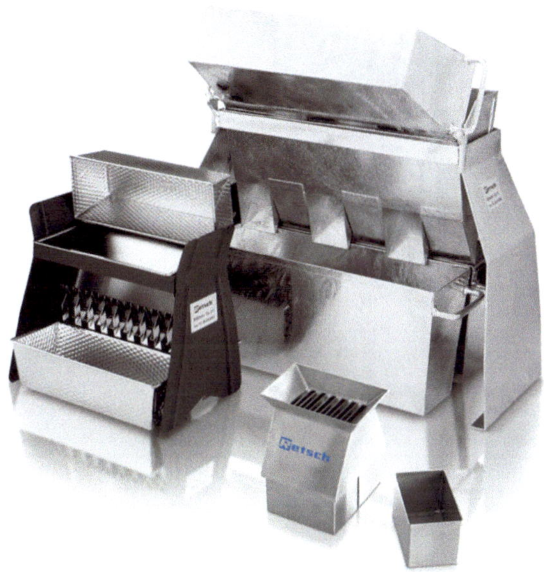

16.5 A Short Real-Life Story

Many years ago, following extremely extensive and thorough groundwork involving raw material tests, simulation of extrusion performance in the laboratory, etc., an order was placed from a company in Europe for the supply of a complete preparation and extrusion plant to produce face bricks. Production started successfully in spring. Seventy per cent of the target output was achieved already after six months, and the quality of the bricks was excellent. However, the production laboratory was still in the process of being built. Everyone was happy—but not for long. Just ahead of Christmas, the problems started to multiply and by spring production was just short of being stopped. The disputes with the client were certainly not pleasant. Only once a neutral laboratory accepted by both parties was introduced, could it be proven that not only the variations in the material moisture content were totally inadmissible, but also the conditions in the quarry changed on a daily basis. Therefore, the ceramic body no longer corresponded by far to the intended operating raw materials. Luckily, both we and the client possessed identical representative retention samples with the aid of which the deviations could be verified. All these problems could have been avoided if there had been a functional laboratory in the first place. But all is well that ends well! Today, this brick factory is producing one of the best face bricks and has established an excellent control procedure by sampling 12 different characteristics of the material. Figure 16.4 is simplified and shows only four of those characteristics.

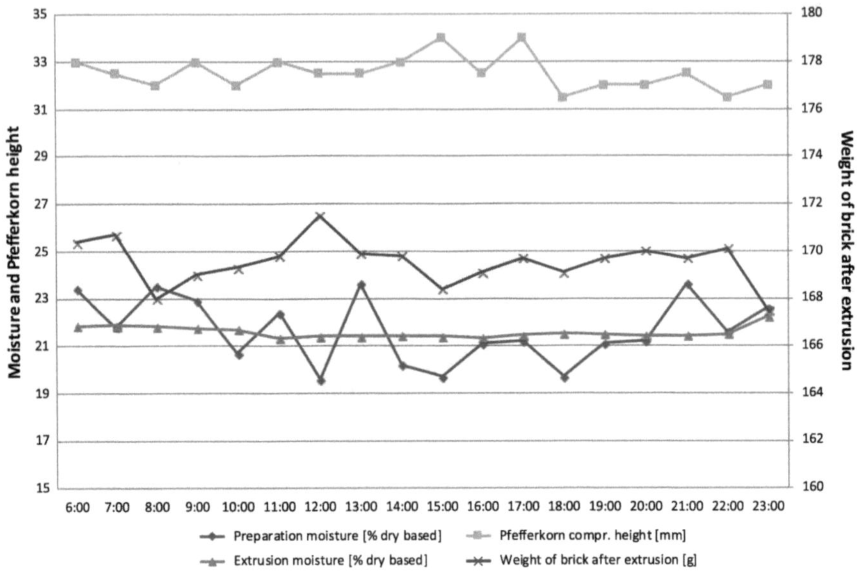

Fig. 16.4 Typical variations of ceramic material at a brick factory

16.6 Methods and Instruments

We are frequently asked which basic equipment we would recommend for a production laboratory. Please bear in mind as you go through the following list of basic equipment:

- This is in fact a checklist which should give you an idea of the basic equipment for a production laboratory.
- For some production facilities, this may appear to be too comprehensive, for others—insufficient.
- The list is based on the assumption that extrusion will be involved in the production process. Other shaping processes such as dry pressing necessitate different test methods and instruments.

In Table 16.1, the basic equipment for the production laboratory in ceramics is listed (Fig. 16.5).

Table 16.1 Basic equipment for production laboratory

Equipment	Test procedure
Air-conditioned room equipped with laboratory table, water basin, tables for the different devices, separate room for meetings and office	General equipment
Computer hardware and software to protocol and evaluate the detected data	General equipment
Tools like hammers, rubber mallets, screwdrivers, spanners, packaging materials, buckets, air-sealed containers of different sizes, containers for toxic material, safety glasses, safety shoes, protective and rubber gloves, cleaning equipment like high-pressure cleaner and all special tools for the instruments	General equipment
Fine roller mill, hammer mill, cross beater mill, ball mill, filter press for laboratory	Crushing equipment for plastic and/or dry material
Laboratory mixer (5–10 l)	Preparation of bodies
Laboratory kneader (5–10 l)	Preparation of bodies
Laboratory vacuum extruder (50 or 80 mm diameter) including pressure heads/dies for the required profiles, foreseen for adequate extrusion pressures	Generating of samples and detecting of extrusion parameters like pressure, temperature, output.
Electric laboratory dryer	Drying of samples for further tests
Electric laboratory kiln	Firing of samples for further tests
Laboratory scale (50 kg, tolerance ±1 g)	Weighing in of different materials to produce samples
Precision scale (1 kg, tolerance ±0.01 g)	Weighing in of different materials for analyses
Moisture analyser with printer	Moisture content
Manual penetrometer Pen check III for laboratory	Plasticity—"deformation pressure"
Luminescence test equipment Capillary rheometer (Capillar Check)	Homogeneity, mixing quality
Rheometer (Capillar Check)	Rheological characteristics of the extruded mass
Shrinkage measurement device	Drying and firing shrinkage of samples
Screening tower, sieve set with 5–8 screens depending on requested fractionation, ultrasonic bath, particle size analyser	Grain size analysis
Combined bending and compression test machine (force >10 kN for bending and >100 kN for compression strength)	Bending and compression strength of dried and fired laboratory samples
Microscope (magnification factor >100)	Structure of mass, detection of impurities
Temperature, shrinkage, moisture—logger	Measurement of real parameters during drying

(continued)

Table 16.1 (continued)

Equipment	Test procedure
PTCR rings	Control of temperature during firing
DTA/DTG instruments	Thermal characteristics and behaviour during firing
Dilatometer	Shrinkage–expansion behaviour during firing

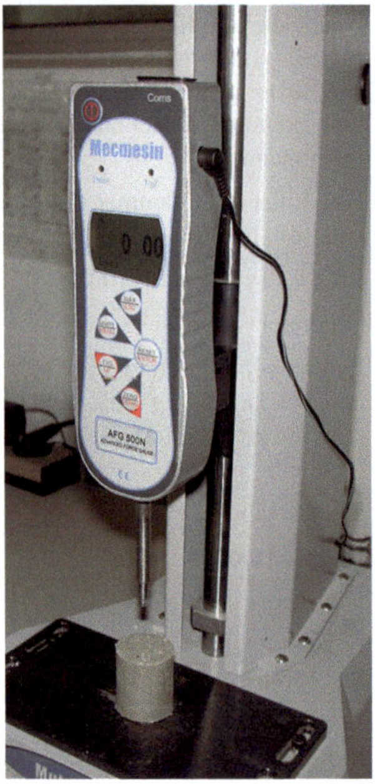

Fig. 16.5 Pen check III to measure the deformation force

16.7 Conclusions

- You, your management and colleagues have to be aware of how important the production laboratory is for your operation.
- Define the hierarchical position of the production laboratory in the organizational structure in view of possible decision-making processes and assuming the worst case scenario.
- Define quality-related features of your products such as moisture content for extrusion, pressure rates, mineralogical structure and also the acceptable range of variations within which no quality defects are to be expected.

- Do not make savings on human resources and technical equipment for the production laboratory. It would be economizing in the wrong place. On the other hand, you should not endeavour to do every test in-house; make use of external services which may be cheaper and better.
- Follow up the relevant statistical literature on the subjects of random samples, sampling, etc.

Bibliography

1. W.G. Cochran, *Sampling Techniques* (Wiley, New York, 1977)
2. DIN 51061-04, Testing of ceramic raw and finished materials—Sampling of ceramic raw materials (2017)
3. K. Göhlert, M. Übel, Test methods for plasticity and extrusion behaviour, in *Extrusion in Ceramics*, ed. by F. Händle (Springer, Berlin, 2007), pp. 347–362
4. M. Probst, W. Bender, in The works laboratory—indispensable for constant product quality. In: ZI 3/2000, 13–32

Chapter 17
Piston Extruders in Ceramics

extrudere (lat.) = to draw or thrust out.

It was in 1619 that John Etherington built the first hand-operated piston extruder for the extrusion of bricks. This was followed by numerous designs.

Piston extruders were, and still are, used not only for the extrusion of ceramic bodies but also find application in many industries for the shaping of plastic materials, or for materials that have been plasticized in a tempering process.

In 1810, Joseph Bramah used the first hydraulically operated piston extruder for the shaping of lead slabs, to produce tubes. Piston extruders are used in the food industry, for the shaping of aluminium profiles, copper alloy and steel, for the extrusion of hard metal, graphite, coal, etc.

In the following, focus will be on the description of hydraulic piston extruders as employed in the ceramic industry and for the extrusion of hard metal, since the application in these two cases is very similar.

17.1 Types of Piston Extruders

Extrusion by means of piston extruders, also called piston presses, is a discontinuous shaping process in which a plasticized material filled into a material container is pushed through a die or perforated plate by a piston. The geometry of the die or perforated plate determines the extruded profiles, called extrudates.

Piston extruders available for use in the ceramic industry can be classified according to the following criteria:

- By the press force (from 1 to 1,000 tonnes),
- By the diameter of the material container (from 10 to 1,000 mm),
- By the type of pressure-generating system (hydraulic or mechanical),
- By the direction of extrusion, (horizontal, vertical or tiltable),
- By the kind of material fed into the material container (bulk material, kneaded material or pre-shaped billets),
- By the method of feeding (manual, semi-automatic, automatic),

F. Händle, *The Art of Ceramic Extrusion*, https://doi.org/10.1007/978-3-031-63471-0_17

- By the type of de-airing, (with/without de-airing),
- By the method of control (Speed-controlled, pressure-controlled, non-controlled),
- By the tempering of the material container (cooled, heated, non-tempered),
- By the method of fixing the tools/dies (mechanical, hydraulic).

In addition, there are a number of the other special criteria and characteristic features. These may relate, for instance, to the difference in the hydraulic system, oil-hydraulic or water-hydraulic versions, or to the degree of sophistication of the control system and which operating data are available online, or whether the cylinder can be changed to suit different diameters, etc.

17.2 Working Principle of a Piston Extruder

The working principle of a piston extruder appears quite simple at first glance. The material (feedstock), in the form of powder, granulate or billets, is filled into the material container and a hydraulically operated piston presses the material through the die, which determines the shape of the finished product (see Fig. 17.1).

At second glance, it is much more complex than it appears. Many small details are necessary to use the machine to produce high-quality products made of ceramics and/or hard metal.

Unlike extrusion with screw extruders, extrusion in piston extruders is a discontinuous process. But depending on the size of the material container and the diameter of

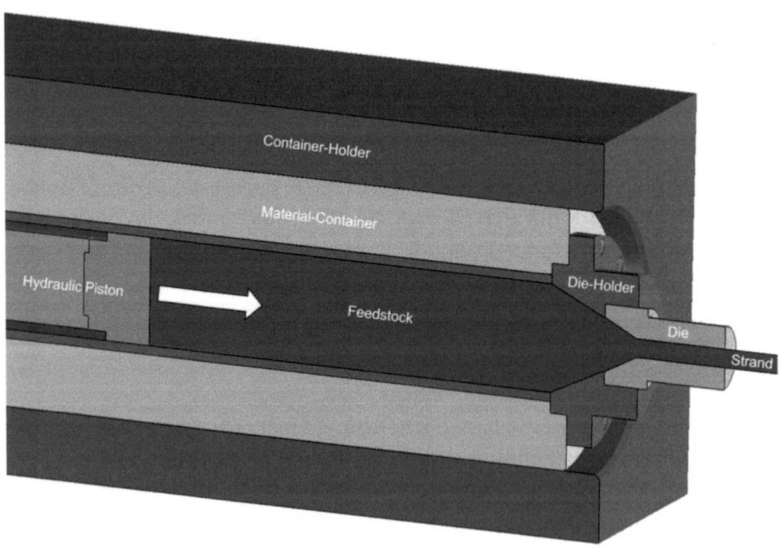

Fig. 17.1 Schema material container during extrusion

the die, the extrusion can operate continuously over hours. As soon as the container is empty, the process has to be stopped and the piston press has to be reloaded.

This is definitely a disadvantage compared to continuous screw extruders, however, piston extruders do have specific advantages that make them essential for the production of special high-quality rods and profiles made out of ceramics and hard metal.

17.3 Advantages of Piston Extruders

If we compare the piston extruder with the screw extruder, we can identify a number of advantages that makes the piston press an efficient alternative.

- High pressure inside the material container.
- The material can be pre-compressed inside the material cylinder by up to 1,400 bars. That guarantees a homogeneous structure of the extrudates.
- Constant extrusion speed.
- Modern piston extruders are equipped with precise control of the extrusion speed, which is very important for the production of small and twisted rods.
- Quick and simple cleaning.
- It takes max. half an hour to clean the material cylinder and the piston head, that is indeed much faster than needed for cleaning a screw extruder.

  Important for the production of small material batches.
- Reduction of laminations.

  Laminations generated by the screw, such as rotation laminations or laminations caused by the hub of the screw, can be avoided with a piston extruder.
- Low wear and contamination.

  Wear and metallic contamination at the cylinder and piston head are marginal compared with the wear and contamination generated by a screw extruder if abrasive material is extruded.
- If low shearing during the shaping process is beneficial for such masses that exhibit "shear thinning" effects", piston extruders can be beneficial compared with screw extruders.

To summarize: Both shaping systems have obvious advantages. Screw extruders are unbeatable for the mass production of products, with similar dimensions and the same feedstock. But if very high extrusion pressures or small batches with different feedstocks are required, piston extruders are advantageous.

17.4 Components of the Piston Extruder

In the following, the main components of modern piston extruders are described (see Fig. 17.2).

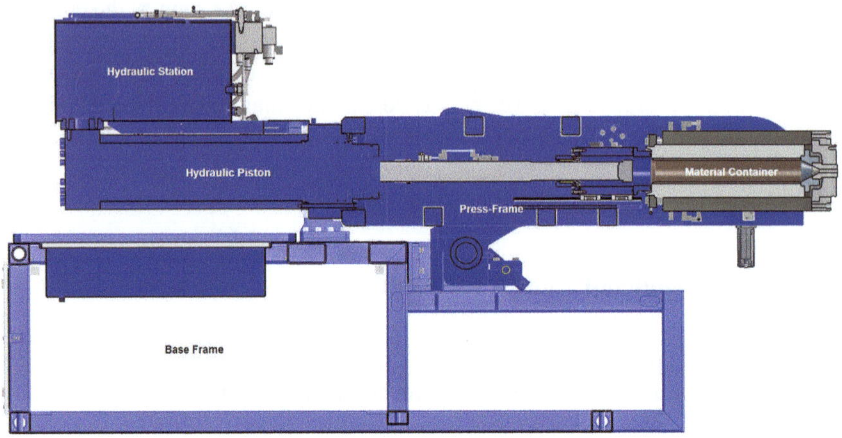

Fig. 17.2 Schema piston extrusion press

17.4.1 Hydraulic System and Hydraulic Piston

The heart of the piston extruder is the hydraulic station. It is responsible for the continuous and constant movement of the piston, and further for a stable extrusion speed.

The drive system of most modern piston presses consists of an energy-efficient servo motor, which regulates the piston speed by changing the rotation speed of the motor. Big hydraulic pumps and proportional valves are no longer state of the art. Efficient servo motors generate no heat and no noise. That makes it possible to install the hydraulic station directly near the piston press without any adverse effects in respect of the relevant "Environmental, Health and Safety Regulations" (EHS).

In combination with a precise linear-encoder, the servo motor is able to move the piston with completely constant speed. An accuracy of 0.001 mm/s is possible. This is remarkable considering that a big hydraulic cylinder with a diameter of 300 mm has to be moved with a slow and constant speed of 0.003 mm/s. This high accuracy is a must especially for small rods or drills with twisted cooling channels.

The slow speed is necessary because of the big difference between the container diameter and the diameter of the product. If a rod with a diameter of 1 mm has to be extruded with a piston press consisting of a container diameter of 120 mm, the shaping ratio is 14,400! That means that the rod is changing the extrusion speed by ±14.4 mm/s, when the piston speed changes by ±0.001 mm/s.

Also important for the hydraulic station is a short distance between the valves and hydraulic cylinder. That guarantees rigid conditions without any influence of the compressible hydraulic oil (see Fig. 17.3).

Most of the piston presses work with a hydraulic pressure of 150–250 bars. But this pressure level should not be compared with the pressure level in the container. This pressure can be much higher, because of the ratio between the hydraulic cylinder

Fig. 17.3 Hydraulic station

and the diameter of the material container, i.e. the feedstock will be compressed with an extremely high pressure to generate a homogeneous extrudate.

17.4.2 Base Frame and Press Frame of the Piston Extruder

The generated force of a piston extruder can be up to 300 tonnes or even higher, depending on the required application. To withstand these high forces, a massive steel frame, called the press frame, is necessary including the hydraulic cylinder on the one side and the material container with the die holder on the other side. The most important function of the press frame is to be absolutely rigid. Every elastic movement of the frame has a negative influence on the extrusion process.

Another requirement for the design of the press frame of the piston extruder is to allow extrusion at different angles between 0 and 90° (see Fig. 17.4). This is necessary because of the different plasticity and different dimensions of the extrudate, to avoid deformations, for example, for small rods with low plasticity.

The counterpart of the press frame is the base frame. It has to take the total load of the hydraulic cylinder including the cylinder for tilting which is mounted inside the base frame.

Fig. 17.4 Press positions

In the past, hydraulic cylinders were installed for the tilting. Nowadays, servo-motor-driven spindles are used to tilt the piston press. This solution is much more energy-efficient and it makes it possible to automatically adjust the press angle with high accuracy.

Components like the vacuum pump, the maintenance unit for compressed air and the electrical cabinets are installed inside the base frame, for easy maintenance.

17.4.3 Material Container

The material container receives the feedstock for one extrusion cycle. The volume of the feedstock that can be loaded into the container depends on the diameter and the length of the container.

Some piston extruders have a fixed container that is mounted on the press frame. Modern piston extruders are split into the container and the container holder. With this solution, more functions are possible, i.e. to optimize the properties of the feedstock.

- Temperature control.

  The temperature of the feedstock has a relevant influence on the extrusion process. In most cases, the force of the piston can be reduced because, on one hand, the material becomes softer at higher temperatures, while the necessary press force is lowered with higher temperature, on the other hand. Also the mass might become sticky and more difficult to handle after extrusion. Depending on the binder, temperature control of the container has to be very precise to generate the optimal temperature for the feedstock. Tempering systems with circulating water at the required temperature are normally used. Even more efficient for heating are electrical heating cuffs.

- Wear protection.
- The material container is the part of the piston press that is in contact with the abrasive material. To avoid wear, the container is made of hardened steel or equipped with shrunk wear bushing made out of hard metal. These wear bushings are polished to guarantee a very smooth surface so as to reduce the friction between material and container.
- Tilting function.
- To realize an ergonomic filling operation, modern piston presses allow tilting of the material container in a low position.

  When the container is tilted back to press position, it is fixed by a locking pin to ensure that the piston is absolutely in line with the container.
- Changing function.
- Depending on the dimensions of the extruded product, it can be necessary to work with different container diameters. Therefore some piston presses are designed to allow easy changeover of the container. This function can also be used to change the container if this becomes worn.

17.4.4 Die Holder

Every manufacturer of powder metallurgy products and quite a few producers in advanced ceramics have their own developed dies. Accordingly, it is important that the die holder is tailor-made for each die system.

To enable a quick change of the dies, two different clamping systems are generally available. Classical clamping by screws or by a modern hydraulic clamping system with bayonet locking, which ensures a quick change of the dies without the need for tools.

The die holder has some further functions. It can be tempered by the heating or cooling system of the material container, which ensures that the material has the right temperature in the die area.

To control the radial pressure of the material at the die, a pressure sensor can be installed inside the conical part of the die holder. This online pressure reading guarantees relevant information regarding the plasticity of the extruded ceramic body. Pressure sensors with an integrated temperature measurement allow control of the temperature.

17.4.5 Vacuum System

Having air inclusions in the extrudate is the worst case for the quality of the product that applies to powder metallurgical as well as to ceramic or hard metal masses. This leads to cracks and bubbles inside the finished product after sintering. To avoid this, the air has to be removed in the material container, before the extrusion process starts.

Therefore a vacuum cap is pressed against the open end of the container, connected to the vacuum pump, which removes the air inside the container (see Fig. 17.5).

As long as an adequate vacuum level is not reached, the piston does not start to move. Without sufficient evacuation, the piston extruder would extrude bodies contaminated with air, i.e. produce extrudates with so-called air laminations. This means that the operative is not allowed to start the extrusion process as long as the evacuation has not been completed.

The vacuum pump is positioned in the base frame of the machine. Good accessibility is needed for proper maintenance, like inspection of the filter or oil changes. The size of the vacuum pump depends on the size of the container. For the production of hard metal it is important that the vacuum pressure is less than −10 mbar during extrusion. Zero leakage from the entire system is essential.

Fig. 17.5 Vacuum cap

17.4.6 PLC and Data Logging

Most modern piston presses are controlled by a high-performance PLC. The PLC is responsible for the automatic movement of the piston and also for all ancillary movements, like changing the tilting angle or turning the material container. Further, the control of the heating or cooling systems, vacuum pump and hydraulic system is controlled by the PLC.

Different sensors like the vacuum sensor, the sensor measuring the radial pressure or the extrusion speed etc. deliver the operational data online during the production cycle. All these data are visible on an HMI (see Fig. 17.6), in form of a touch panel. For the operative, it is important to see operational data online during production.

The most important values are visible in diagrams directly on the HMI. According to the requirements of the Industry 4.0-orientated standards, it is important that all these values can be stored as a CSV file and/or can be sent directly to SAP or similar systems.

All operational data for different products can be stored as recipes, which makes it comfortable and safe for the operative to adjust the piston press automatically and ensures that the different processes are repeatable. In the recipes, the type of graphite trays for the cutting and handling system can be selected and stored to ensure a fully automatic process downstream of the extrusion station.

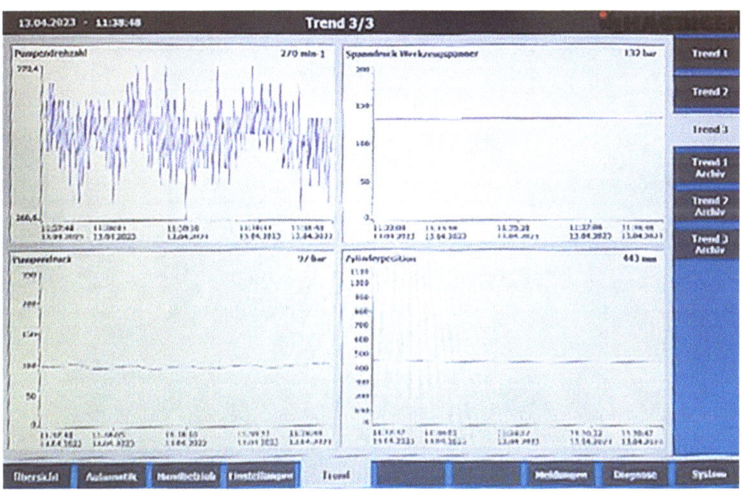

Fig. 17.6 Visualization of process data

17.5 Auxiliary Equipment

The extrusion system is a sub-system in a complex production system consisting of sub-systems like storage of the material including additives, dosing, preparation, extrusion, cutting and handling, drying, sintering, quality control, etc. to mention only the most important sub-systems. Two relevant sub-systems, namely, handling/ cutting and automation are described in the following.

17.5.1 Handling and Cutting Device

In the past, the extrudates used to be cut off manually by an operative and placed on a graphite tray for transport of the products into a dryer or a sintering furnace. Specially for sensitive products like deep-hole drills, twisted rods with cooling channels made of hard metal or cutting knifes, manual manipulation can be problematic, because deformations can be not avoided.

Therefore, fully automatic systems were developed that eliminated the need for manual operations (see Fig. 17.7).

Fig. 17.7 Extrusion press with handling and cutting device

The benefits of a repeatable production without deformation of the extrudate are obvious.

Cutting should be performed by means of a flying knife or a wire-cutting device. The flying knife has the advantage that the end of the extrudate does not exhibit any damage.

Adhesive, sticky extrudates are another factor influencing the handling. To overcome the problem caused by the adhesive surfaces of the extrudates, products can be transported on a kind of air cushion.

Vibration is an alternative if the air cushion is not an option because of uncontrolled drying effects.

Especially for very small products like micro-drills, it is a must to transport the products by a special conveyor, as otherwise these rods would bend and deform.

With this solution, it is possible to produce extrudates with small diameters with a length up to 1600 mm.

The handling and cutting devices can be integrated near the piston extruder or they can be designed as a standalone unit on a carriage. The integrated solution is preferable, because it is a more stable solution and the accuracy is higher.

Both cutting and handling solutions can be used not only for piston presses but also for screw extruder systems.

17.5.2 Automation of a Piston Extruder

Today, it is possible to operate the piston extruder fully automatically during a complete extrusion cycle. Even the filling of the material container can be automated, including pre-compression of the bulk material and evacuation.

A six-axis robot is used for automatic feeding and handling the trays (see Fig. 17.8). This robot can feed the piston extruder with powder or billets and sort the sintering trays in a carriage or a fixed shelf.

The function of the operative is reduced to the transport of the carriage or shelf to the dryer or furnace, including a first stochastic quality control of the products.

The automation not only improves productivity, it also helps to relieve the operative from heavy and sometimes unhealthy work. One operative can supervise several extrusion lines.

Automated quality checks during the extrusion process are possible with high-performance measuring systems, which can, for example, control the pitch of a drill or check if the mandrels of the cooling channels are still working. These preventive quality controls during and after extrusion are very beneficial, because mistakes can be detected immediately, avoiding follow-on waste costs.

Fig. 17.8 2 pcs. extrusion presses with automatic robot handling

17.6 Further Developments

The fundamental characteristic of a piston extruder distinguishing it from a screw extruder is its intermittent mode of operation, from which all its essential benefits but also disadvantages derive. There are clear benefits of the piston extruder as described in Sect. 17.3.

Back in the nineteenth century, this initiated research work to develop so-called continuous piston extruders.

In order to reduce charging times, a second receiver is freshly loaded with material and placed in position (see Fig. 17.9).

As empirical tests have proven, the structure of the extrudate of the last filling is not the same as the structure of the extrudate of the following filling. In practice, often the products extruded during this short transition period have to be discarded. To avoid this effect, piston extruders have been developed that guarantee near-constant extrusion and avoid the effects during the transition period (see Fig. 17.9).

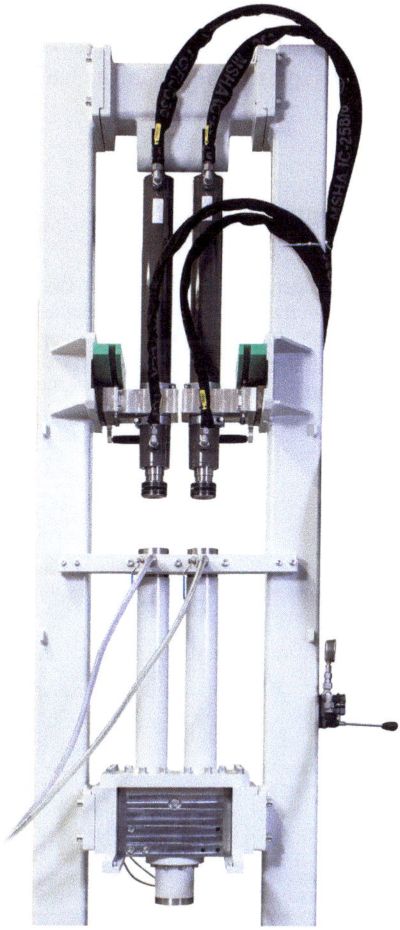

Fig. 17.9 Continuous piston extrusion press

17.7 Conclusions

- Discontinuous working piston presses are beneficial for the extrusion of smaller batches of ceramic and hard metal parts, whilst continuous extrusion with screw extruders is beneficial for larger quantities.
- Piston presses are the preferred extrusion units if extrusion with extremely high pressures up to 1,400 bar is required.
- A servo-motor-controlled hydraulic pump is energy-efficient and silent in operation.
- Modern piston presses can operate automatically, including the filling, cutting and handling operations.
- Data logging of all process data gives good feedback for process and quality control.

- Piston presses are useful when small batches of ceramic mass have to be extruded because the cleaning of a piston extruder takes less than half an hour.
- Piston presses do not create laminations like rotating screw extruders.

Bibliography

1. W. Bender, H. Böger: "A Short History of the Extruder in Ceramics", In: F. Händle (Ed.): Extrusion in Ceramics, Berlin, Heidelberg 2007
2. G. Doll, F. Händle, F. Spiessberger: "Piston Extruders" In: F. Händle (Ed.): Extrusion in Ceramics, Berlin, Heidelberg 2007
3. See the recommended books in the bibliography in Chapter 1.
4. See the websites of Haginger Maschinenbau GmbH & Co.KG (www.haginger.com) ECT-KEMA GmbH (www.ect-kema.de)

Chapter 18
Does that Exist—The Optimal Extruder?

The gravest doubt is the father of insight.
—Max Weber

Abstract The final selection of the optimal extruder is only possible by comparing the relevant and specific technical properties of the alternative extrusion units. A checklist with the relevant properties and the preoccupation with the rich literature on decision theory should be helpful.

Keywords Optimal extruder · Characteristics

18.1 A Sad Story

Let us just imagine that you have decided in favour of a specific extruder offered by a certain manufacturer, but no order has been placed as yet. The price looks attractive, the salesman is a nice guy, you had been invited to an excellent lunch, and the prospects of utilizing his holiday apartment in Tuscany are quite appealing. Just before the order is about to be placed, your boss enters your office and asks "Are you absolutely certain that you have opted for the optimal extruder suitable for our requirements?" You try to explain to your superior that there is no optimal extruder as such. This seems to leave him fairly unimpressed and he replies, "I would like to discuss this offer with you and all other persons involved next Friday whereby I expect you to prove to me in detail why the extruder suggested by you will meet all our requirements". To end the story: the meeting with the boss and your colleagues goes completely against your expectations. The extruder you selected fails to fulfil the requirements in four essential respects—and on top of that, the holidays in Tuscany came to nothing.

F. Händle, *The Art of Ceramic Extrusion*, https://doi.org/10.1007/978-3-031-63471-0_18

18.2 Questions and Criteria

What are the criteria for the selection of the optimal extruder? (Fig. 18.1).

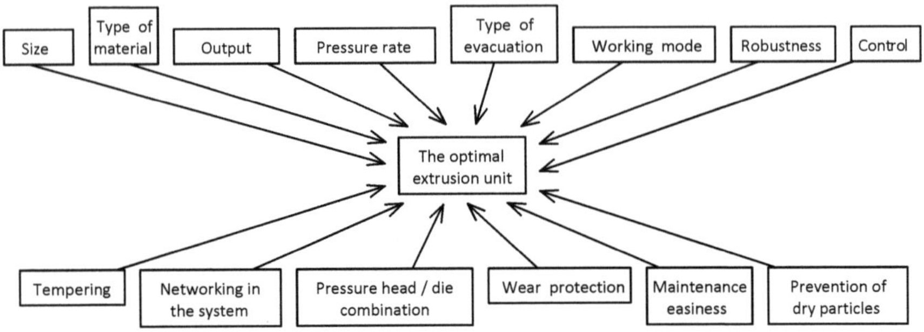

Fig. 18.1 Optimal extrusion unit

The following simplified checklist presupposes that you are familiar with the essential properties of your material, that you know the targeted outputs of all extrudates, also the working times and other relevant boundary conditions. We also assume that the "optimal" extruder can only be selected by a comparison between the viable alternatives. However, this would require that the details of the alternative options are known and that the manufacturers provide transparent offers—which by far are not always the case.

The checklist—as an example—is set out for comparison of three extruder units. In order to provide a better understanding, the first column has already been filled in with entries designated to a fictitious manufacturer A (Table 18.1).

Table 18.1 How to find the optimal extruder by comparison

	Essential characteristics of A, B and C	A	B	C
1	Do you need a stainless steel version, if yes is the complete system stainless steel and rustproof?	Yes		
2	What is the nominal torque of the gearboxes of the extruder/pug sealer and who is the maker?	60,000 Nm		
3	How many kW can be installed at extruder/pug sealer?	75 (kW) 37 (kW)		
4	What is the dynamic load of the axial bearing of the extruder?	70 t		
5	What is the most suitable diameter for my product and why? Shaping degree?	250 (mm)		
6	What is the maximum admissible radial pressure of the extruder?	130 bar		
7	What is the roughness (Ra) of the auger surface?	1–2 Ra		
8	What is the ratio of the auger (L/D)?	6.32		
9	Is the complete system—including auger and vacuum chamber—cooled? If, yes to which temperature?	Yes −5 °C		
10	Can the different zones of the extrusion unit be cooled separately?	Yes		
11	Has the supplier already delivered systems for my product?	Yes		
12	Is the supplier capable of simulating the extrusion process by CFX methods?	Yes		
13	Has the supplier a laboratory background with proven scientific systems to check the rheological characteristics of our body?	Yes		
14	What kind of wear protection can the supplier offer?	PTA and HVOF		
15	Data of the installed vacuum pump? (pumping speed and ultimate pressure)	30 m^3/h, 20 mbar		
16	How is a constant dosing of the extruder guaranteed?	Describe		
17	Which equipment is intended for body preparation?	Describe		
18	What kind of control system and how many online sensors are foreseen?	Describe		

18.3 Optimal Extruder = Optimal Extrusion?

Assuming you have found the optimal extruder to suit your requirements. Does this also mean that optimal extrusion conditions are met? Not at all! In the course of your laborious search for the optimal extruder, you had to assume certain boundary conditions in respect of the characteristics of the body, such as constant homogeneity and plasticity—which are by no means always present.

Let us summarize again the essentials for adequate conditions as follows:

- The correct optimized "recipe" for the required extrudate
- Knowledge of the rheological parameters of the body through scientific laboratory tests

- Constant dosing of the combined de-airing unit
- No dry agglomerations before and during extrusion
- High constant and sufficient vacuum.
- Constant and optimized plasticity of the body through controlled temperature, sufficient ageing, shear energy and optimized homogenization
- Perfect tempering of the body before and during extrusion
- Optimized wear protection of the wearing parts
- Optimized auger and liner geometry
- State-of-the-art online control of all relevant extrusion parameters.

18.4 Conclusions

- Before deciding in favour of, or against, an extruder goes through all significant features in a systematic manner, compare and evaluate these features in view of your requirements.
- Do not rely on your own experience. Never forget what Richard Feymann said: "The first principle is that you must not fool yourself and you are the easiest person to fool."
- Make sure that all boundary conditions which are the basis for an optimum extrusion are met.

Bibliography

1. F. Händle (ed.) *Extrusion in Ceramics* (Springer, Heidelberg, Berlin, 2007)
2. J.S. Reed, *Principles of Ceramic Processing* (Wiley, New York, 1995)

Chapter 19
Additives for Extrusion

"There are many recipes for the perfect chocolate cake, good ones and bad ones, but only one right one."
—Rinaldo, chef and baker from Rovereto

19.1 Objectives for the Use of Additives

The basic requirement for shaping by means of extrusion is a ceramic body with sufficient plasticity. A definition of the property "plasticity" is provided in Chap. 6, so it is not necessary to include one here. In the processing of ceramic bodies that contain sufficient quantities of clay or kaolin, the clay minerals contained in these raw materials bring plastic properties with them, providing the water content is within a certain range. Moreover, heavy clay products, for example (backing bricks, clay pavers, roofing tiles), stoneware pipes, clay slugs for porcelain tableware and porcelain insulators, and refractory chamotte bricks can be extruded very effectively solely on the basis of the plasticity of the clays or kaolins in the body formulation.

In technical ceramics, however, bodies are extruded that consist largely or completely of entirely non-plastic raw materials like, for example, alumina, zirconia, silicon carbide, silicon nitride or aluminium nitride. Therefore, as part of the body production and preparation, certain additives must be added to obtain a body with optimum plasticity and extrusion properties.

With the use of these additives, the following objectives are pursued:

- Extrusion or shaping of the required extrudate cross-section with minimal pressure.
- Water retention even at relatively high pressures (avoidance of water being "squeezed out" of the ceramic body).
- High conveying efficiency of the auger.
- Guarantee of maximized bond in the body and therefore avoidance of cracks during the shaping and handling of the extrudates.

F. Händle, *The Art of Ceramic Extrusion*, https://doi.org/10.1007/978-3-031-63471-0_19

- Profile stability of the column after shaping as well as during transport or handling (no deformation).
- Minimized friction between body and steel inside the extruder, in the pressure head and die, and accordingly minimization of wear on the machine components that come into contact with the body.
- Good behaviour of the extrudates during drying.

Moreover, the additives should not have an adverse effect on the process and the quality of the product in the downstream technological steps like drying and sintering. This includes the requirement for the additives to break down without any problems, namely during sintering or one of the debinding processes upstream of sintering, and ideally escape completely as gas from the ceramic microstructure. Therefore they are mostly organic substances that only have a temporary effect up to the dried extrudate.

This restriction is necessary because small quantities of inorganic materials are added to many materials in high-performance ceramics, e.g. sintering aids, grain growth inhibitors or substances to prevent lattice transformation during heating (stabilizers). With the use of such additives, however, no influence on extrusion or the properties of the extrudates is intended.

Moreover, the following remarks will concentrate only on substance systems that contain water as fluid phase. In technical ceramics, sometimes water-free material systems are extruded. The fluid phase can consist, for example, of an organic solvent if the ceramic particles are not permitted to come into contact with water in order to avoid unwanted reactions (e.g. in the case of aluminium nitride).

Furthermore, in this chapter, the numerous thermoplastic additives that are used, for example, in injection moulding are not addressed either. The same applies for the waxy additives that are used amongst other things in the extrusion of carbides, whereby in such cases the auger and the barrel segments of the extruder are heated to defined temperatures usually above room temperature.

19.2 Types of Additives

19.2.1 Plasticizers

The most important plasticizers for bodies for the production of oxide and non-oxide ceramic materials are cellulose ether. In operation practice, the term "cellulose" is often used, it is, however, not technically correct. Cellulose is the polysaccharide most commonly found in nature, it being a primary structural component of plants. The cellulose molecule is regular, rigid chain without branches, which consists of up to tens of thousands of anhydroglucose units. Every glucose unit contains three reactive hydroxyl groups (OH groups) that form a network of hydrogen bridge bonds. Therefore cellulose is not soluble in water and therefore not usable in this form as a plasticizer (Fig. 19.1).

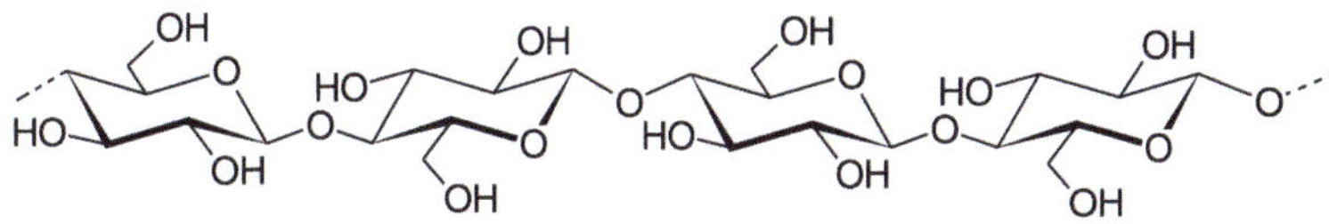

Fig. 19.1 Part of a cellulose molecule (glucose units with reactive hydroxyl groups)

In a chemical process (esterification), the reactive OH groups of the cellulose are substituted partly or completely by functional groups, and consequently water-soluble cellulose ether is formed, which can be used very effectively for plasticizing ceramic extrusion bodies.

Depending on the process management of the chemical reaction and the chemicals used, the cellulose scaffold can carry different functional side groups (e.g. methyl-, hydroxypropyl-, hydroxyethyl groups) and in addition the degree of substitution can be different (Fig. 19.2).

Accordingly, a large number of very different cellulose ethers is available on the market, which are sold in powder form (see Table 19.1).

As already mentioned, the cellulose ethers are soluble in water. The aqueous solutions can have very different viscosity, which depends on the type of cellulose ether, the degree of substitution, the size of the macromolecules and the concentration. Figure 19.3 shows the viscosity of an aqueous HEMC solution as a function of the shear rate. At low concentration, NEWTONIAN behaviour is observed, while at higher concentration the solution shows structural viscosity, i.e. the viscosity decreases with increasing shear stress.

Figure 19.4 shows an SEM image of a cellulose ether in which the macromolecules can be seen. A mix of the ceramic powder (e.g. Al_2O_3), the cellulose ether and water forms a plastic body. The macromolecules of the cellulose ether bond the

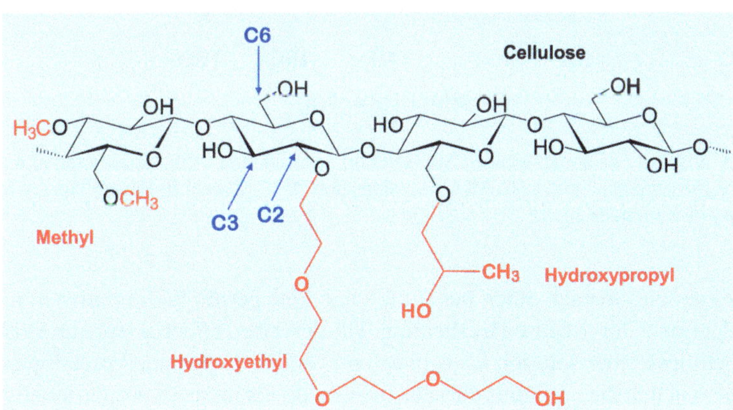

Fig. 19.2 Examples of functional groups on the cellulose scaffold after esterification. *Source* Roland Bayer: Paper "Cellulose Ether Binder" at Freiberg University of Resources on 17.05.2018

Table 19.1 Important types of cellulose ethers. The cellulose ethers used predominantly in ceramics are shown in bold type

Name of the cellulose ether	Abbreviation
Methyl cellulose	MC
Hydroxypropyl methyl cellulose	HPMC
Hydroxyethyl methyl cellulose	HEMC
Hydroxyethyl cellulose	HEC
Carboxymethyl cellulose	CMC
Ethyl cellulose	EC
Ethyl methyl cellulose	EMC
Ethyl hydroxyethyl cellulose	EHEC
Hydroxypropyl cellulose	HPC

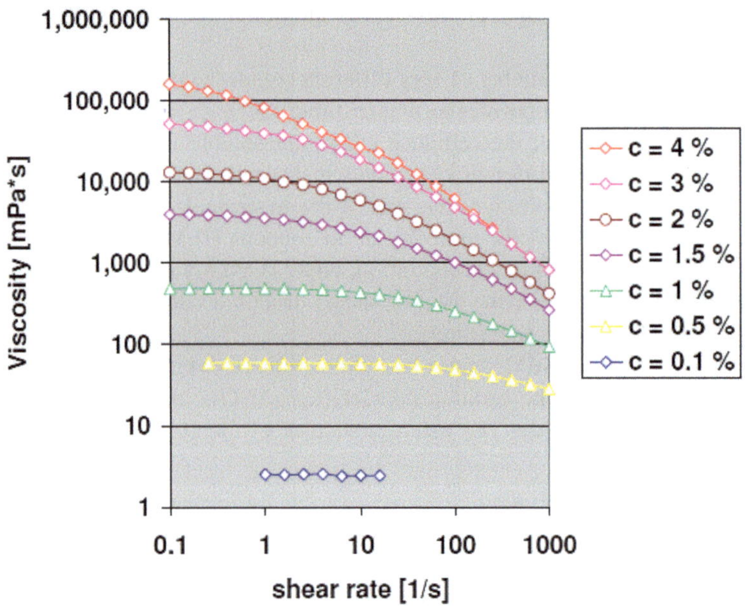

Fig. 19.3 Viscosity of an aqueous HEMC solution as a function of the concentration c and the shear rate, measurement with a HAAKE viscosimeter at 20 °C (*Source* R. Bayer: Paper at Freiberg University of Resources 2018)

ceramic particles to each other, but at the same time permit their relative movement on application of force caused by shearing. The described effect of structural viscosity of the cellulose ether solution leads in ceramic bodies to so-called pseudoplasticity. It is shown in that the pressures during extrusion do not increase proportionally to the shear rate. This is referred to as shear thinning, which, for example, is very positive during extrusion of honeycomb bodies with very thin walls measuring 150 μm. The high shear rates up to around 1,000 s^{-1} in the thin slots of the die result in a very

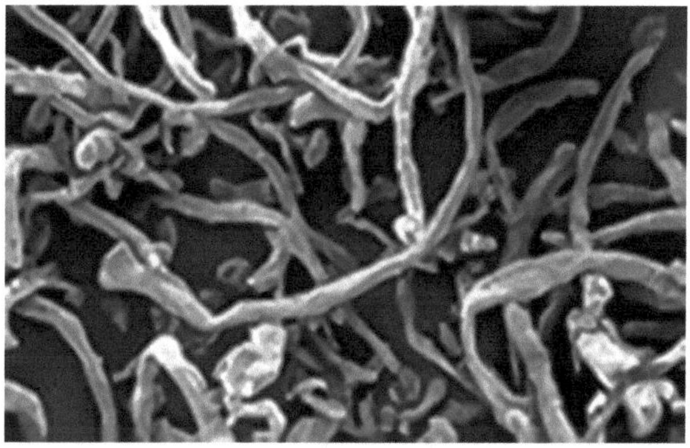

Fig. 19.4 SEM image of macromolecules of a cellulose ether

low viscosity of the body, and accordingly trouble-free and crack-free extrusion of the honeycomb structure is possible.

For many cellulose ether solutions, in the case of increased temperatures, a thermoreversible gelling results, the cause of which is a structural change of the macromolecules. The result of this gelling is that the ceramic body loses its plasticity, changes into a brittle state and therefore becomes no longer extrudable. The temperature at which this gelling occurs is also called "gel point". The gel point temperature measured in a pure cellulose ether solution is, however, not identical to that of a ceramic body that contains the corresponding cellulose ether. Besides the type, the concentration of the cellulose ether in the ceramic body has an influence on its gel point temperature. Figure 19.5 shows the influence of the temperature and the concentration of the cellulose ether "Methocel®A4M" on the pressure for extrusion of an Al_2O_3 body. It can be seen that with increasing concentration of the cellulose ether, the required pressure becomes steadily lower, but this also lowers the gelling point, which manifests itself in a sudden increase in the pressure.

In the meantime, several cellulose ethers have been developed that exhibit sufficiently high gelling points for the practical operation. In addition, many auger extruders are cooled so that the danger of an excessive body heating is almost excluded.

Cellulose ethers are specified according to the viscosity of a 2-% aqueous solution. In most cases, this viscosity is included in the product name as a number without measurement units. So there is, for example, a methyl cellulose ether MC 3000 (around 3,000 mPa s), HPMC types 4M (around 4,000 mPa s), 15M (around 15,000 mPa s) or even 100M (around 100,000 mPa s). The higher the viscosity, the larger the macromolecules are in the same cellulose ether type.

The manufacturers sell the cellulose ethers usually under certain brand names, e.g. Benecel™, Culminal™ (Ashland), WALOCEL™ (Dow) or Tylose® (Shin-Etsu). In operating practice, these brand names are often found in the internal formulations.

During the sintering process, the cellulose ethers are supposed to break down completely and without residue. That is especially necessary for ceramic products with properties that have to meet high requirements. One example from the field

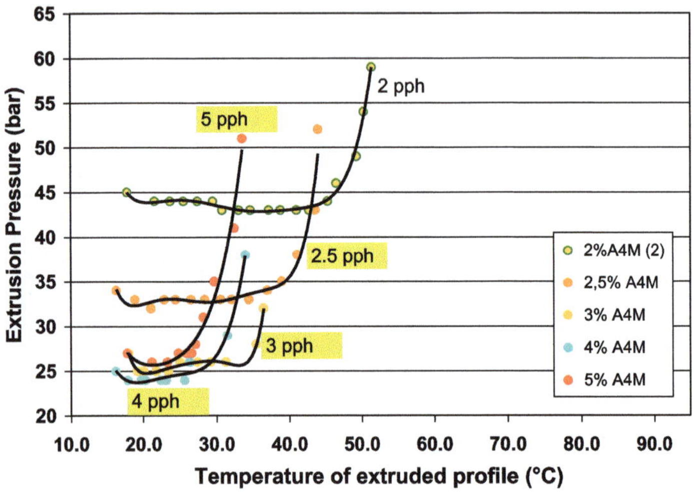

Fig. 19.5 Pressure as a function of the extrudate temperature (Methocel® 4M added in different quantities to a Al_2O_3 body) (*Source* R. Bayer, M. Lang, cfi/Ber. DKG 89(2012)5)

is explained in the following: The objective was to develop a body for an Al_2O_3 component that had to exhibit very high bending strength. In all tests, the same Al_2O_3 powder was used and the formulation was always unchanged. Only the type, degree of substitution and the molar mass of the cellulose ether were varied. As a result, it was established that the cellulose ether used has a very significant influence on the porosity and strength of the component sintered at 1750 °C, although it should have actually broken down in the sintering process and escaped in the form of gas from the ceramic microstructure (Fig. 19.6).

Besides cellulose ethers, in individual cases starch is also used for plasticizing ceramic bodies. Starch is a polysaccharide with the formula $(C_6H_{10}O_5)n$ and extracted from maize, potatoes, wheat or other plants. Starches are cheaper in price than cellulose ethers, their effect in respect of plasticization is, however, very limited. A body with the addition of starch only shows a certain flowability at strong shearing. Consequently, extrusion of a body with exclusively starch additives is not possible with the use of a piston extruder. For the shaping of such a body by means of an auger extruder, this must be able to introduce high shear energy into the body.

19.2.2 Lubricants

Apart from the plasticizers, lubricants are almost always added to the extrusion bodies. These have two functions:

(a) Reduction of the external friction, i.e. the friction between the extrusion body and the metal parts of the extruder, especially in the die.

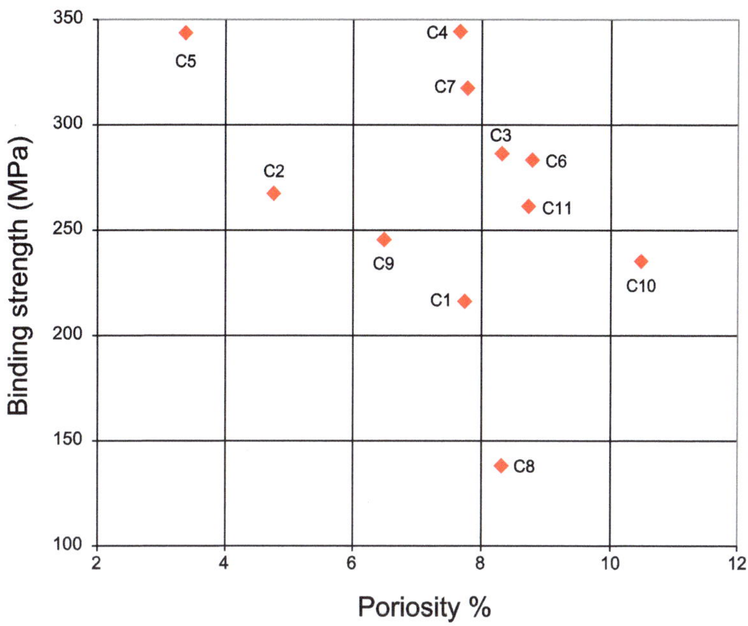

Fig. 19.6 Influence of various cellulose ethers (C1 to C11) on the bending strength and porosity of sintered alumina. The highest value of strength is 2.5 times larger than the smallest value. Formulation: 84.8 mass% Al_2O_3, 2.3 mass% CE, 0.9 mass% lubricant, 12.0 mass% water

(b) Reduction of the internal friction, that is the friction between the raw material particles.

With the addition of suitable lubricants, the following effects are achieved:

- Reduction of the pressure.
- Increase in the specific throughput rate respectively increase in the transport efficiency of the auger.
- Homogenization of the rate of the column exit over the cross-section of the die.
- Reduction of the body's tendency to stick, e.g. in the inlet of the extruder or in the de airing chamber.

As lubricants for extrusion bodies, mainly the following substances are used:

- Polyethylene glycol (PEG).
- Polyethylene oxide (PEO).
- Glycerine.
- Fatty acids (e.g. oleic acid, stearic acid).

Polyethylene glycol (PEG) and polyethylene oxide (PEO) belong to the same chemical group of substances, they are polyethers that differ in their chain length and the molecular mass. The general molecular formula is $C_{2n}H_{4n+2}O_{n+1}$.

Polyethylene glycols have a molecular mass in the range of 200 to 35,000 g/mol and can be liquid, pastose or solid. At a temperature of 20 °C, low-molecular PEG in the region of 200 to 400 g/mol are liquid, in the region from 400 to 3,000 g/mol, they are pastose and at even higher molecular masses, they are present as solids that can be sold as powders or flakes. The molecular mass of polyethylene oxides is larger than 35,000 g/mol, that is why they are always in solid state at room temperature.

In respect of their use as additives for ceramic water-based bodies, it is especially important that both polyethylene glycols and polyethylene oxides are water soluble. From a PEG with molecular weight of 8,000 g/mol, a 50-% aqueous solution can still be prepared.

Moreover, PEG and PEO are non-toxic, which is why these substances are also used in cosmetics, foodstuffs and pharmaceutical products. Thanks to the exceptionally low toxicity values and good skin compatibility, their use in the preparation and processing of ceramic bodies is simplified, since no special working and health regulations apply for the handling of these substances.

The quantity of lubricants added is usually under 1.0 mass percent active substance relative to the dry substance of the ceramic raw materials. The specification in "active substance" is important because polyethylene glycols are often used as aqueous solutions and in the calculation of the formulation, the concentration of the solution must be taken into account. In practice, addition in the form of a liquid or solution during the mixing and kneading processes is preferred because relatively fast and homogeneous distribution and incorporation in the ceramic body are possible as a result.

Another lubricant used is glycerine, which is a sugar alcohol with the molecular formula $C_3H_8O_3$. It can be mixed with water in any ratio, that is why it is added with the entire or part of the mixing water to the body during the mixing or kneading process. An especially notable property of the glycerine is its water binding capability. As a result, the stability of the water films around the ceramic particles is apparently improved and segregation of the solid-water on the application of pressure to the ceramic body (in the pressure head and die) is reduced.

Numerous preparations offered on the market as lubricants for ceramic bodies contain fatty acids. Fatty acids are monocarboxylic acids with a usually unbranched carbon chain. Short- and medium-chain fatty acids contain up to 14 carbon atoms and are readily water soluble. As lubricants in ceramic bodies, mainly long-chain fatty acids like, for example, oleic acid (molecular formula $C_{17}H_{34}O_2$) or stearic acid (molecular formula $C_{17}H_{36}O_2$) are used. At a temperature of 20 °C, oleic acid is liquid, stearic acid, on the other hand, is solid. Both, however, are almost water insoluble.

The stearic acid is commercially supplied in the form of flakes or beads, with these products often containing other fatty acids like, for example, palmitic acid. These flakes (or beads) feel waxy and are very soft. On addition to the body mix, it is necessary to ensure that in the preparation machine (intensive mixer or kneader) sufficient shear energy is introduced to grind the several-millimetres-large solid particles and distribute these homogeneously. Experience has shown that this task can be performed effectively by intensive mixers.

On the market, there are also fatty acids that are made emulsifiable based on reaction, e.g. with ethylene oxide, as a result of which they can be better incorporated into the body.

19.2.3 Additives to Influence the pH Value

In many cases, small quantities of an organic acid, e.g. formic acid or acetic acid, are added to a ceramic extrusion body. Formic acid is the simplest carboxylic acid with the molecular formula CH_2O_2, and acetic acid, too, (molecular formula $C_2H_4O_2$) is a carboxylic acid. As a result the dispersity of the particles (e.g. Al_2O_3) and consequently the flowability of the plastic body are increased. To understand this effect, it is necessary to examine more closely the surfaces of the ceramic particles and the interaction with the surrounding medium (usually water). Oxide particles (e.g. Al_2O_3, ZrO_2, TiO_2, etc.) possess at the surface non-saturated bonds because the atoms show no complete coordination of their next neighbours as in in the interior of a crystal lattice. As soon as the surface of such a particle comes into contact with water, it becomes hydrated, i.e. ions from the liquid phase are deposited on the surface. Both H^+ ions and OH^- ions are adsorbed, although this depends on the pH value of the aqueous substance system. There is a characteristic pH value at which a particle together with its hydration shell is outwardly neutral. This pH value is called the isoelectric point (IEP). For α-Al_2O_3, this point lies at pH 9 to 9.5, for zirconia ZrO_2, the IEP is specified at pH 4 to 5.

If you add an acid (e.g. acetic acid) into a Al_2O_3 body mixed with water, the pH value is reduced and all particles together with the hydrate shell then have a positive charge (Fig. 19.7). As like charges repel, the particles can move more easily against each other, the flowability of the body is improved.

It should, however, be noted that with the analysis, the processes speculated on here have been highly simplified. The ceramic material systems are very complex electrochemical and -kinetic processes on micro- and nanoscale that are described extensively and in-depth in the relevant scientific literature.

19.3 Technological Aspects of the Use of Additives

Additives are constituents in the formulations of an extrusion body and accordingly they are added to the body as part of its preparation. Prior to them working with additives, it is imperative that machine operatives carefully study the safety data sheets for the substances and strictly follow the safety instructions. In many cases, protective goggles, safety gloves and, in some cases, protective masks must be worn if the substances are, for example, carcinogenic, acidic or caustic.

As it is often a matter of very small proportions (sometimes less than 1 mass%), particular attention should be given to the mixing process. If an additive is available

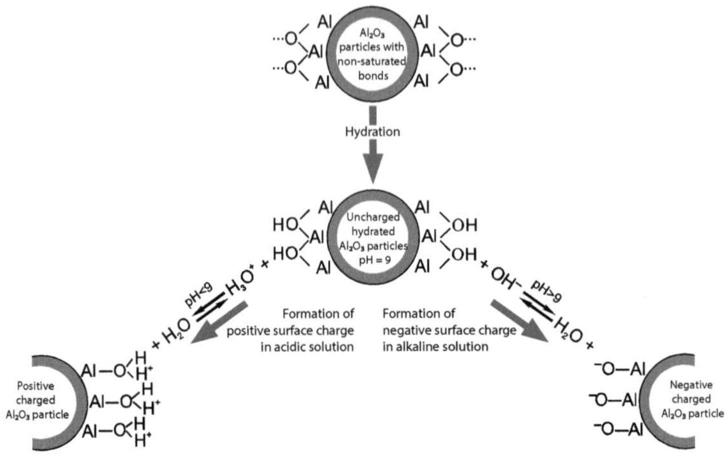

Fig. 19.7 Formation of hydrated Al_2O_3 particle surfaces and their reactions in acid and alkaline solution (schematic) (*Source* J.G. Heinrich: Lecture manuscript "Technology of Ceramics", TU Clausthal 2005)

in solid form, it is filled with other powder raw materials into the mixing vessel (or optionally into the kneading container) and mixed without the addition of any liquid until a homogeneous powder has been formed. The more intensive the mixing process is, the shorter the time necessary for it. Particularly effective are so-called intensive mixers with a rotating mixing vessel in which one or several tools (agitators) operating at high speed ensure very high relative movement of the powder particles. Here, the escape of dust into the environment must be avoided with appropriate measures, e.g. seals, extraction, enclosures.

It is best to add liquid additives when the mixing or kneading tools are already running. In many mixers, it is possible to introduce liquids by means of spray nozzles. If additives are water soluble, first an additive solution is prepared in water that is then fed into the mixing or kneading machine.

Among additives, cellulose ethers generally make up the largest proportion in the extrusion bodies. These are more or less fine powders that are generally weighed with the other raw material powders and mixed. With the addition of water into the mixer, the cellulose ethers absorb the largest part of the water, leading them to swell. In the course of this process, a plastic body is formed. The swelling process of the cellulose ethers requires a certain time. That is why, prior to processing in the extruder, the mixed body is stored in closed containers where possible for a certain period (e.g. 24 h). This ageing can be performed in special rooms in which a certain atmosphere (high relative humidity, low temperature) is maintained with dedicated technical equipment.

Another possibility is to dissolve the cellulose ether separately in water and let it swell it in advance. In respect of the water temperature, this is adapted to the optimum solubility conditions of the cellulose ether used. For that reason, specialized agitation equipment with heated containers is very suitable. A gel-like mass is obtained that can then be incorporated in powder ceramic raw materials in another mixing or kneading machine.

This technology requires, however, that the required quantity of water is sufficiently high in relation to the mass of the cellulose ether. In many cases, however, this is not the case.

19.4 Conclusions

- Most extrusion bodies for the production of technical ceramics (oxide and non-oxide ceramics) contain certain substances, so-called additives, in small quantities, to render them in a flowable plastic state and to enable the trouble-free shaping of the extrudates at minimized pressures and with low wear of the extruder components that come into contact with the body.
- The most important additives are plasticizers (e.g. cellulose ether), lubricants (e.g. polyethylene glycol) and substances that can selectively change the pH value (e.g. carboxylic acids).
- Every manufacturer of extruded technical ceramics has their own "secret recipe" in respect of the additives used. This recipe is often a result of a relatively long historical development process.
- The optimum additives in respect of type and additive quantity can only be determined with systematic tests in practice-oriented laboratory trials, the expertise of the employees being of great importance in the development of new products and new materials.

Bibliography

1. Kollenberg, W. (Hrsg.): Technische Keramik. 3. Auflage. Vulkan Verlag Essen 2018, Kapitel 4.3. Organische Additive. S. 408–413
2. M. Hölzgen, P. Quirmbach, „Additives in Extrusion" In: F. Händle (Ed.) Extrusion in Ceramics, Berlin, Heidelberg 2007
3. Knarr, M.; Bayer, R.: The shear dependence of the methylcellulose gelation phenomena in aqueous solution an in ceramic paste. Carbohydrate Polymers 111 (2014) S. 80–88
4. Bayer, R.: Cellulose Ether Binders. Synthesis and Chemical Properties. Vortrag TU Bergakademie Freiberg 17.05.2018
5. Bayer, R; Lang, M.: Corrosion-resistant Alumina Tubes through Extrusion, Part 1. Cfi/Ber. DKG 89(2012)5, S. 115–121
6. Bayer, R.: Die Steuerung der Prozesseigenschaften bei der Extrusion keramischer Massen über die Pseudoplastizität der Celluloseether. Plenarvortrag auf der DKG-Jahrestagung Hermsdorf 23.03.2010

7. Heinrich, J.G.: Technologie der Keramik. Vorlesungsmanuskript TU Clausthal 2005, Kapitel 4 "Formgebung", S. 96–169
8. Gauckler: Herstellung von Keramik, Band 1. Vorlesung an der ETH Zürich 2001. Kapitel 7.1. "Keramische Partikel in flüssigen Medien" (S. 93–110)

Chapter 20
Additive Manufacturing by Extrusion in Ceramics

3D printing is the most radical idea since the invention of the Internet
—Dr. Anthony Atala

20.1 Introduction

Since its invention by Chuck Hull in 1981, additive manufacturing (AM), also referred to as 3D printing, has been hyped as a megatrend among the different methods for shaping components. By far—but not only—for components with complex geometry, with the stereolithographic processes currently dominating.

Components are printed from different materials such as plastics of all types, metal, carbon, synthetic resin, concrete as well as from ceramic materials.

At the relevant trade fairs and in the flood of literature, it can be difficult to comprehend the developments in this sector. Increasing speeds in the generation of objects, larger dimensions, improved surfaces and the production of exceptionally filigree structures have led to new applications constantly opening up.

In the meantime, a profusion of different processes and variations can be subsumed under the term additive manufacturing. But in all cases, AM differs from other shaping processes like casting, dry pressing, etc. in that on the basis of CAD design data, a material—in powder, plastic or liquid form—is deposited in layers onto a platform and then stabilized by means of various hardening processes.

In the draft standard DIN EN ISO/ASTM 52,900:2018 and the VDI Guideline 3405, more than a dozen different processes are listed. Among these processes, listed under the abbreviation SEAM is "Screw Extrusion Additive Manufacturing".

As not only screw extruders, but piston presses can also be used for extrusion within the scope of AM, we opine that the term AM by extrusion is appropriate.

Whereas in common AM processes, discrete volumes of different consistency are deposited, key feature of the processes developed for ceramic and silicate materials is the continuous deposition of layers. Depending on the application, layers

F. Händle, *The Art of Ceramic Extrusion*, https://doi.org/10.1007/978-3-031-63471-0_20

with a diverse range of cross-sections can be extruded thanks to the widely variable geometry of the nozzle.

Among the steadily growing range of AM processes, stereolithographic processes in numerous variations are easily dominant. In contrast, the AM extrusion processes are positioned in defined niches to meet specific requirements by harnessing the advantages of the AM extrusion process.

20.2 Theoretical Principles of Extrusion

For better understanding, readers are referred to the theoretical principles of extrusion addressed in Chaps. 3 and 4. Depending on the configuration—see 21.3, relevant variations of the operational data like, for instance, axial and radial pressures, can result.

20.3 Configurations

In the initial tests within the scope of AM for the production of ceramic components, the basic configuration consisted of a stationary extruder and a mobile nozzle connected to the extruder by means of a hose.

The objective was to determine and evaluate which pressure losses and temperature changes result with different hose lengths and hose cross-sections (see bibliography).

Basically, the following configurations are possible and have been realized in practice:

- Stationary extruder versus mobile extruder.
- Stationary platform versus mobile platform.
- Nozzle fixed to the extruder, nozzle separate from the extruder.

These configurations can also be combined with each other.

For better understanding, a few examples are given in the following.

In the construction industry, for printing houses and walls, the configuration often consists of a mobile extruder/printhead which is constantly fed by a stationary cement pump and is controlled by a portal crane.

For the manufacture of small objects, the configuration often consists of a platform mobile in two horizontal dimensions onto which the layers are deposited and an extruder with a nozzle that is only mobile in the vertical plane.

Other examples are described in Sects. 20.5.1 and 20.5.2.

20.4 Hardening of the Green Extrudates

For hardening of the plastic ceramic or silicate columns/layers, essentially the following processes are known or under development.

- **Chemical hardening**

  The process developed by 3D Minerals is described under 20.5.2.

- **Hardening by fusion-melting**

  The fusion melting common for the printing of metal components and subsequent hardening is also possible for AM by extrusion, but is still under development for columns with relatively large diameters/volumes.

- **Drying**

  Leather-hard drying at the column exit is sufficient in many cases to avoid deformation during deposition of the columns.

- **Thermoplastic extrusion**

  Thermoplastic extrusion of feedstocks consisting of ceramic and plastic materials is a common process in ceramic technology and is likely to develop into one of the most interesting alternatives for AM by extrusion: see Section 20.5.3.

- **Hydraulic hardening**

  Hydraulic hardening of ceramic or cementitious feedstocks is a common process.
  The hardening time can be shortened with the addition of accelerators or lengthened with retarders; see Section 20.5.1

20.5 AM by Extrusion—examples in Development and Practice

20.5.1 Development by ECT-KEMA GmbH/TU Dresden

In the following, a system is described that consists of an extruder with limited mobility with nozzles fixed to the extruder, combined with a stationary platform. This system is used for making components from cement, can, however, also be used to make ceramic objects. In the vertical Z axis, the extruder is moveable by 1 500 mm, in the horizontal X and Y axes by 1 000 mm. The moving speed is a maximum of 9 m/min. Depending on the screw speed, throughput rates up to 4.8 m^3 green concrete/hour are generated. The length of the movement paths and the movement speeds depend on the design of the handling equipment.

With this system, 3D-printable concrete with a maximum particle size of 16 mm can be processed. It is used to produce load-bearing solid concrete walls measuring up to 150 mm in width (Figs. 20.1, 20.2, 20.3).

20.5.1.1 System Components

Figure 20.4 visualizes the 3D printhead developed by ECT-KEMA in collaboration with TUD Dresden University of Technology. It consists of two main units, that is the extrusion and shaping modules. The extrusion module meters the concrete depending on the printing speed and the cross-section to be extruded. The shaping module shapes the column of concrete as required and stabilizes this until it reaches green stability.

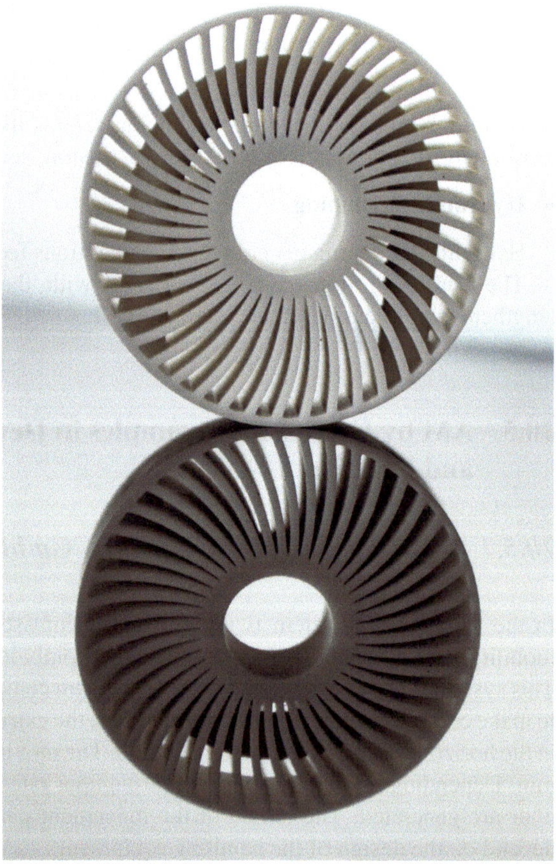

Fig. 20.1 Filigree components printed at 3D CERAM

Fig. 20.2 Tests with different hose lengths

At the standardized fast coupler (1), the feed pipe system with the concrete pump is connected in order to feed the printhead with concrete. In the reservoir (2), the concrete is intermediately stored and mixed thoroughly again by the agitator (3) following the pumping process and then fed to the screw. The screw extruder (5) meters the quantity of concrete needed for deposition as a function of the printing speed and cross-sectional area of the column profile to be extruded. During the extrusion process, the vibrator (4) vibrates the concrete. This keeps the concrete in a sheared state, as a result of which stiffening is prevented and the extrusion process facilitated. The screw extruder is constructed as a hollow shaft, which permits the integration of an interior vibrator that can vibrate the screw. With the integration of the vibrator, the conveying efficiency of the concrete can be increased considerably. The reservoir and the screw extruder are decoupled from the machine frame by rubber elements, so that the vibrations generated by the vibrator are not transferred to other components.

The extrusion and shaping modules are connected by a quick coupler (6), which is integrated as a bayonet fitting. The rotary actuator (7) adjusts the orientation of the outlet nozzle corresponding to the direction of printing. In the cross-sectional transition (8), the basic rectangular column shape is produced. The cutter (9) and

Fig. 20.3 Right-angled wall section with a wall thickness of 150 mm and a layer height of 50 mm

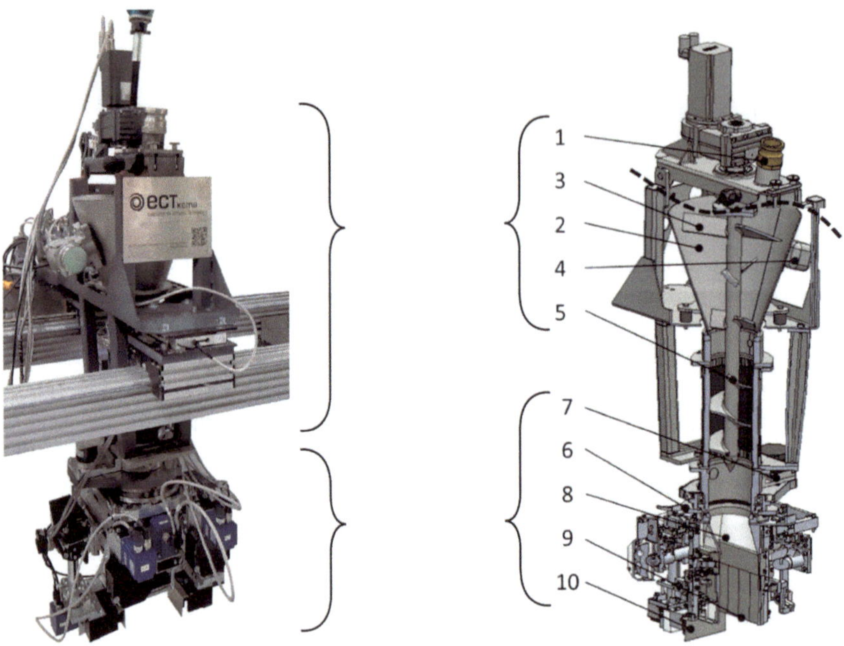

Fig. 20.4 Printhead setup. 1 Fast coupler. 2 Reservoir. 3 Agitator. 4 Vibrator. 5 Screw extruder. Quick hitch7 Rotary actuator. 8 Cross-section transition. 9 Cutter. 10 Shaping baffles

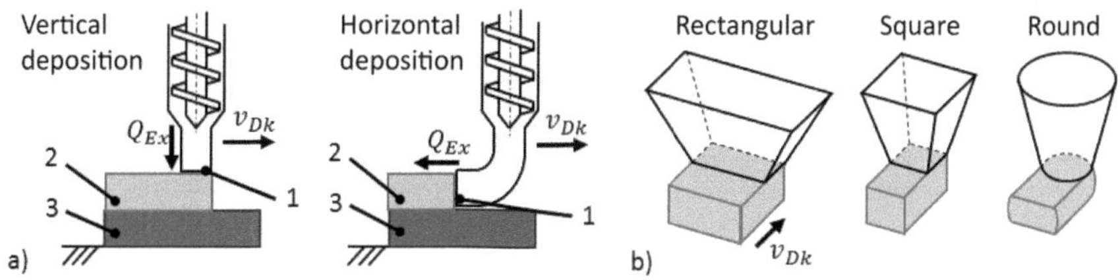

Fig. 20.5 Design of the discharge outlet. Q_{Ex} Extrudate flow; v_{Dk} Printhead speed; 1 Discharge outlet; 2 Current layer; 3 Previously deposited layer. **a** Direction of discharge, **b** Outlet geometry

the shaping baffles (10) model the building elements and shape the final column geometry (Figs. 20.6, 20.7, 20.8, 20.9, 20.10, 20.11, 20.12, 20.13).

20.5.1.2 Nozzle/Discharge Outlet

At the nozzle discharge outlet of the shaping module, the material exits the printhead and the column with the required geometry is deposited. Figure 20.5a shows two different discharge directions of the extrudate Q_{Ex} relative to the printing direction v_{Dk}.

In the case of vertical deposition (Fig. 20.5a, left), the discharge outlet (1) is above the layer to be printed (2) and the flow of the extrudate is perpendicular to the direction of printing. This principle corresponds to the standard option in 3D printers. The form of the discharge outlet is arbitrary, see Fig. 20.5b. The distance between the discharge outlet and the previously deposited layer (3) allows easy adjustment of the current layer height, the deposition pressure improves the bond between the layers [9, 11]. In the case of horizontal deposition (Fig. 20.5a, right), the discharge outlet is located inside the layer to be printed, the material is deposited in the opposite direction to the printing direction. The layer cross-sections are rectangular or square, as shown in Fig. 20.5b [10]. The production of smooth and pore-free layer surfaces is supported by the diversion of thc cxtrudatc flow insidc thc shaping modulc.

The object to be produced is generally composed of repeating individual elements, for example end-elements, corners, curves or T-connections. An increasing complexity of the geometry elements, like variation of the column width or sharp edges, necessitate additional cutters or shaping baffles for their production cf. Figure 20.4. [Sect. 20.5.1.1].

20.5.1.3 Control of the System

The frame of the extrusion module is mounted by means of two load cells on the manipulator. Based on the measured force, the filling level of the hopper is measured and the material feed controlled. With the arrangement of the four bending beams

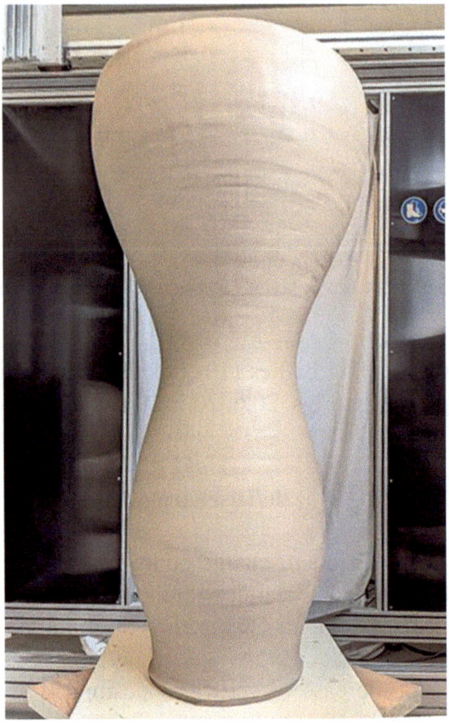

Fig. 20.6 Stoneware. With a wall thickness of 10 mm, the part weighs 40 kg, printing time of 2 h

per load cell, any dynamic influences caused by the movement of the manipulator are offset.

For monitoring of the printing parameters and the condition of the concrete, the system can be equipped with pressure and temperature sensors. Accordingly, variations in the material can be detected and evaluated.

To ensure a constant exit geometry, the deposited column is scanned by a profile scanner. Based on the height and width together with the speed of the handling equipment, the actual discharge volume can be calculated. This is compared with the theoretical discharge volume, and the speed of the screw is regulated correspondingly. This ensures that the geometry of the column is within specified tolerances. All measured values can be logged, enabling continuous quality assurance.

20.5.2 3D Minerals Development

3D Minerals: 3D ceramic printing for special components in small series.

3D Minerals designs and develops ceramic 3D printers equipped with an innovative and patented paste extrusion technology: Slurry Deposition Modelling. This innovation enables 3D printing of large components without interruption of the

Fig. 20.7 Pure alumina (99.7% AI_2O_3). With a wall thickness of 4 mm, and a height of 400 mm, printing time of 25 min, organics < 1%

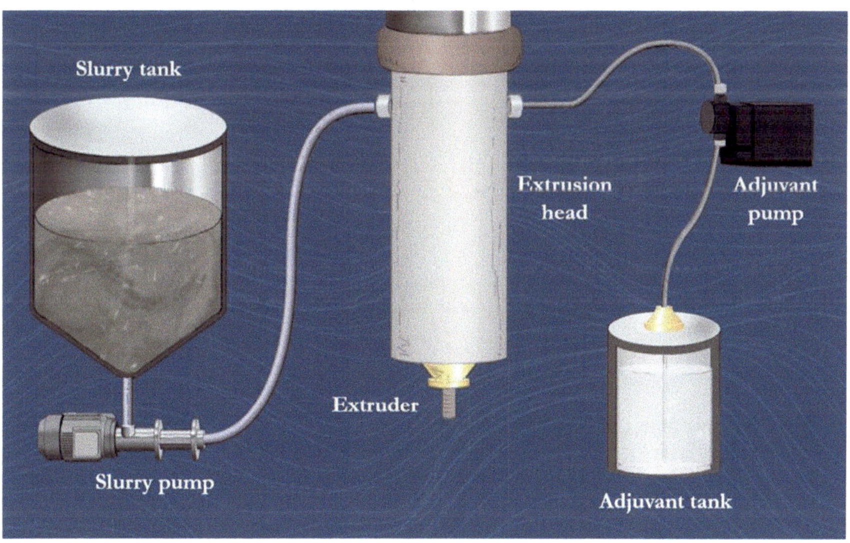

Fig. 20.8 Diagram showing slurry deposition modelling technology

Fig. 20.9 Designer object printed by robotic printer

Fig. 20.10 Extrusion head of rotary printer

Fig. 20.11 Robotic Ceramic 3D Printer
Printing size: Depending on the robot size. For complex parts thanks to the six axes. For automated production of different parts. With rotary table for 7 axes

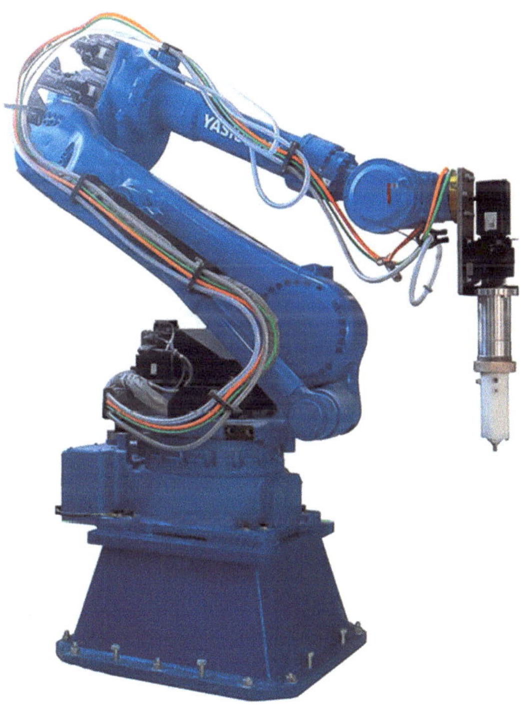

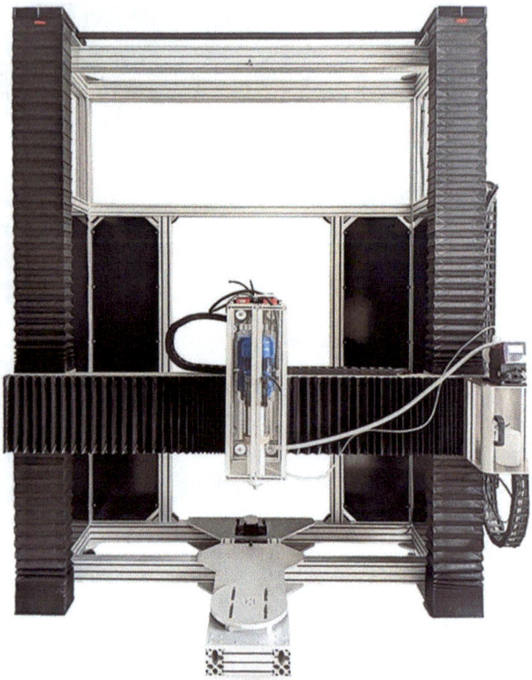

Fig. 20.12 3D Printer with rotary table Standard size: H = 2 000 mm Ø = 1 500 mm. Optimal for printing round parts. Equipped for finishing inside and/or outside during printing.

printing process, and with simple and fast implementation of the ceramic paste to be extruded.

This innovation offers many advantages over the robocasting technique, which uses a plastic mass. The wet process enables continuous production without air bubbles—an important factor, especially for manufacturing large components.

The slurry deposition modelling process patent relies on the rapid reaction between a slurry and additives that convert the liquid into an extrudable paste.

The thread laid down during the printing process also has the property of hardening in a controlled manner over time. All this makes it possible to vary the mechanical properties of the body during printing, adapting it to customer requirements.

The equipment is easy to use as the slurry container can be easily placed away from the extrusion head. Industrial users can also use their slip directly.

S.D.M. technology can also print any type of raw material, including customer raw materials.

This technology enables technical ceramics to be printed with a very low level of additives (around 1%), thus avoiding the debinding stage!

In respect of the different applications, 3D Minerals offers printers with different configurations.

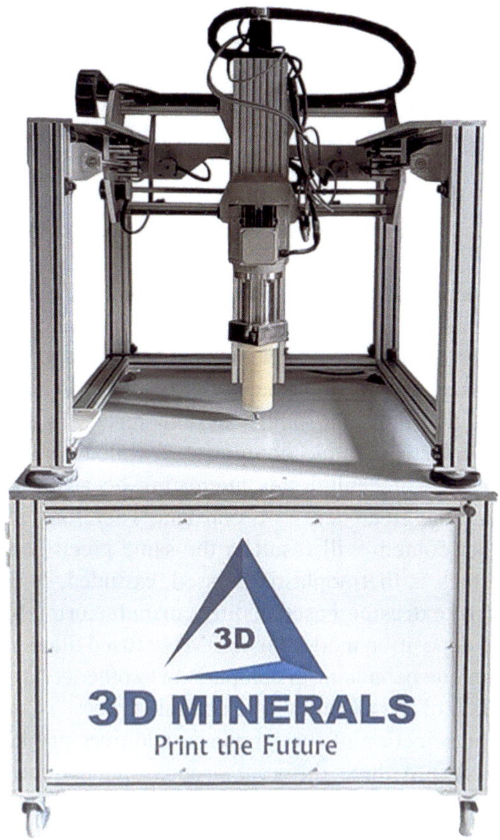

Fig. 20.13 Cartesian 3D Printer Standard printing size: 500 × 500 × 500 mm. Modular 3D Printer: S.D.M or P.D.M Technology. Mainly used for technical ceramics. Designed for studies and prototyping

20.6 Three Machines, Three Sizes

These three machines can print traditional ceramics (stoneware, porcelain, etc.) and technical ceramics (alumina, cordierite, etc.).

3D printing in general and its ceramic applications eliminate the use of moulds, which are difficult to manufacture, expensive and often not recyclable. With 3D printers, individual components or small series can be produced directly from a digital file. Also, some shapes can only be made using this printing technology. We are modifying existing materials so that they can be printed with acceptable quality and speed.

20.6.1 AM by Thermoplastic Extrusion

Frank Clemens, High Performance Ceramics Laboratory—Empa, Swiss Federal Laboratories for Material Science and Technology, Dübendorf, Switzerland.

20.6.1.1 Thermoplastic Shaping of Ceramics

Thermoplastic ceramic shaping methods have been widely explored for different applications. For these methods, a thermoplastic binder system is mixed with ceramic powder to achieve sufficient plasticization, above the melting point of the thermoplastic, during the shaping process. Cooling below the melting temperature of the thermoplastic binder system results in solid green parts. The mixture of thermoplastic-ceramic composite is called ceramic feedstock. The thermoplastic processing of ceramics is interesting because, independent of the shaping technology, the green density is constant. Therefore, a feedstock with the same ceramic powder content will result in the same green density independent of whether the feedstock is thermoplastic-pressed, extruded, injection moulded or 3D printed by material extrusion-based additive manufacturing MEx-AM methods, also known as fused deposition modelling (FDM) or fused filament fabrication (FFF) [12, 13]. This is a unique behaviour in comparison to other ceramic shaping methods. A schematic diagram of the thermoplastic ceramic process is shown in Fig. 20.14.

The selection of thermoplastic additives needed for ceramic feedstocks can be divided into three different groups, namely the backbone polymer, plasticizers and surfactants. The thermoplastic backbone polymer endows the feedstock with

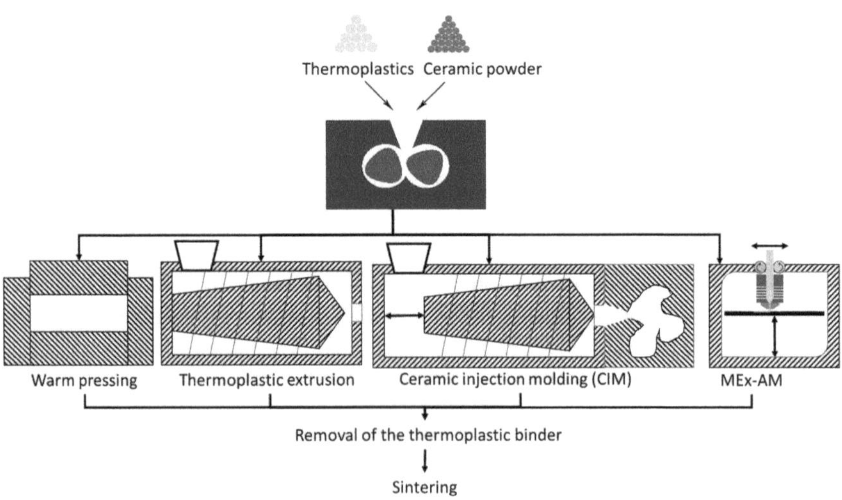

Fig. 20.14 Schematic showing thermoplastic ceramic processing

mechanical strength at room temperature. Plasticizers are used to reduce the viscosity of the ultra-high-filled thermoplastic-ceramic composite, and surfactants are needed to generate good adhesion of the thermoplastic additives on the surface of the ceramic powder. In addition, some researchers have recommended so called tackifiers for MEx-AM process to achieve better fusing between the different layers [14, 15].

20.6.1.2 Thermoplastic-Material-Extrusion-Based AM Methods

The production of plastic components via the layer-by-layer deposition of extruded filaments goes back to the development by S. Scott Crump [16, 17]. Back in the 1990s, the process was further developed for metallic and ceramic materials [18–20]. On account of the patent situation, the process has only become more widespread recently.

A thermoplastic ceramic feedstock needs to be heated above the melting temperature of the polymeric binder to achieve the necessary plasticity for the shaping process. All printheads are therefore equipped with a thermistor and thermocouple near the nozzle to enable regulation of the processing temperature and adjustment of the flow behaviour of the feedstocks. The feedstock needs to be pushed through a nozzle by pressure. The way the pressure is applied can vary and thus we can distinguish between four different printheads, as shown in Fig. 20.15.

The working principle of piston extruder heads is similar to direct ink writing where a pressure gas or plunger will push the material through a nozzle or needle. Scheithauer et al. introduced a printhead utilizing a heatable cartridge designed for printing a ceramic paste made from paraffin and beeswax to print droplets of the

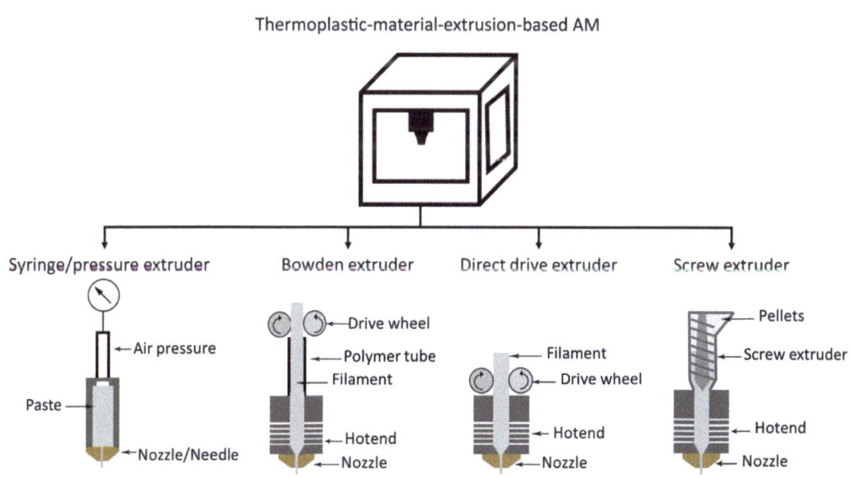

Fig. 20.15 Four different extrusion heads that have been already used for thermoplastic-material-extrusion-based additive manufacturing of ceramics (From left to right: piston, Bowden, direct drive and auger extruder printing heads)

material [20]. Cartridge-based printers require further enhancements like printing precision and resolution when compared to other well-established thermoplastic MEX-AM printheads.

Instead of a paste, ceramic filaments can be used for 3D printing, too. These filaments are typically produced by feeding pellets into a hopper and extruding them with a single auger extruder through a die and finally winding them onto a spool. To enable the use of commercial printheads, designed for polymers, filaments with a diameter of 1.75 or 2.85 mm are typically fabricated and are already commercially available from different companies. For precise adjustment of the filament dimensions, the diameter is online-controlled based on a laser system and the speed of the winding unit is automatically adjusted. A 3D CAD model is translated into a G-code for the printer by slicer software. In addition to the coordinates to print the 3D object, the G-code includes printing parameters such as layer height, speed, infill, temperatures for the extruder head and the printing bed, fan speed and others. To calculate the correct infill for the printing, the filament diameter is an important parameter and therefore the correct adjustment of the filament diameter should not be underestimated. To print these filaments, either Bowden or a direct-drive extruder can be selected. The main difference between a Bowden and a direct-drive printhead is the position of the drive wheels that push the filament into the heating zone of the printhead. In a Bowden-extruder-based printhead, the driving wheels that transport the filament into the printhead are close to the filament spool [22]. Between the driving wheels and the printhead, the filaments are guided in a plastic tube to avoid buckling. This makes the printhead very light and more precise for quick changes in print directions and fast printing speeds. For such print heads, flexible and strong filaments are required and low friction is needed to transport the filament through the polymer tube. When ultra-high-filled thermoplastics, as in the case of ceramic filaments, are used, the thermoplastics become more brittle, and friction inside the polymer tube increases. However, Fig. 20.16a shows that even with ultra-high filler loading, flexibility of the ceramic filament can be still achieved. Nowadays, direct-drive extruder heads are most frequently used (Fig. 20.16b) [23–25]. In a direct-drive printhead, the drive wheels are mounted on the printhead above the heat sink.

Compared to alternative printheads, filament extruders offer exceptional printing quality and surface finish. They are user-friendly to operate and material changeovers are straightforward. However, filament blockages within the extruder head can

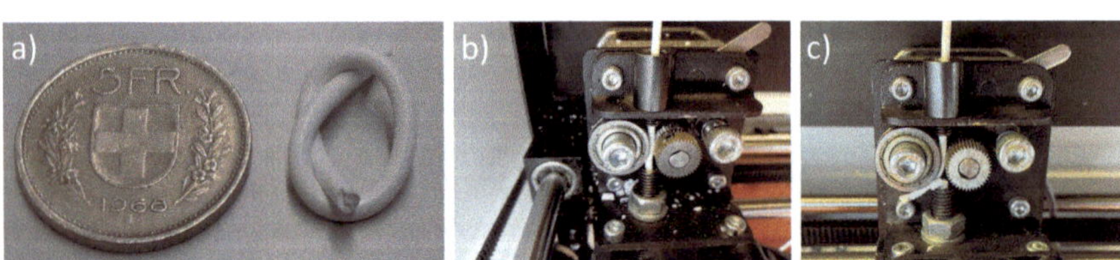

Fig. 20.16 **a** Flexible ceramic filament with 47.5 vol% zirconia content, **b** Zirconia filament in a direct drive printhead, **c** buckling of zirconia filament at higher printing temperature

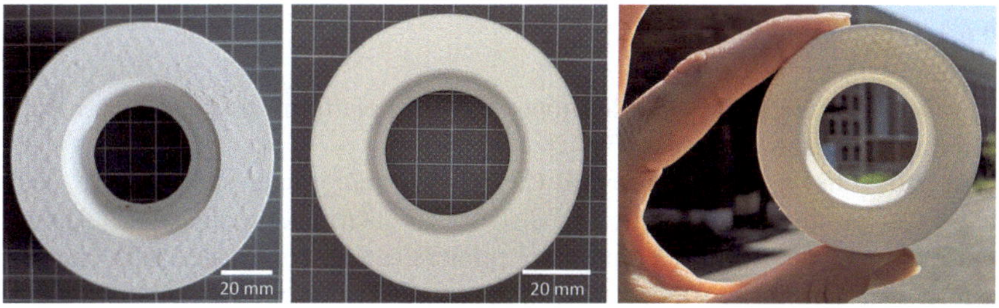

Fig. 20.17 Printed and sintered Al_2O_3 pull roll for polymer filament production lines. The ceramic roll was printed with a consumer-grade printer (Ender 5 pro) and commercial filaments supplied by Fabru GmbH

impede material flow through the nozzle, a common issue known as clogging. While printing at elevated temperatures can alleviate clogging, overheating can cause the filament to buckle, as higher temperatures reduce the stiffness of the solid filament beyond the heat sink (Fig. 20.16c).

As mentioned above, ceramic parts can be printed using commercial ceramic filaments. The price of the filaments is still significantly higher, almost double that of ceramic pellets for ceramic injection moulding. However, cheap filament printers can be used to achieve ceramic parts with good resolution. Figure 20.17. shows a ceramic pull roll for a polymer filament production line. The life of such a ceramic roll is ten times longer than that of a steel roll, and the honeycomb infill pattern means that the amount of ceramic filament used in production can be reduced.

To avoid the high cost of ceramic filaments, pellet extruder heads have become popular in recent years. The extruder head consists of a hopper and a screw [26], similar to extruders and injection moulding machines. The pellets are filled in a hopper and the auger transports them to the entry of the printing nozzle. Inside the extruder, the material is heated gradually before the heating block, and depending on the length of the screw extruder we can differentiate between a transport, melting and compression zone. In auger-extruder-based printheads, length constraints often pose challenges with regard to distinguishing between the various zones. Owing to the short length of the auger, preheating near the hopper can lead to pellet bridging, resulting in decreased material output or potential disruptions of the printing process. Back in 2005, Bellini et al. already used an auger extruder to print ceramic parts and they called the process mini extruder deposition (MED) [27]. The time taken to print large parts varies depending on the screw diameter, which affects the material output per unit of time. However, challenges remain with material retraction during printhead movement to new locations, leading to oozing where molten feedstock forms thin strands between old and new print areas. There are also issues with calibrating material output for different printing speeds, which can lead to under- or over-extrusion phenomena [28].

20.6.1.3 Printing Platforms

Over the years, engineers have developed different platforms that can be generally used for the MEx-AM process (Fig. 20.18.).

As shown in Fig. 20.18, polar and Cartesian system printer platforms are available. Polar printers utilize angle and distance to determine the location of points within the plane. It is notable that these printers can be constructed with just two motors, resulting in very low energy consumption. In contrast, Cartesian-based printer platforms employ the X, Y, and Z axes to define printing coordinates.

In the realm of Cartesian-based printer platforms, various configurations exist depending on how the movement is distributed between the printhead and the print bed. For instance, in some models, the print bed moves in the X–Y direction while the print head moves along the Z-axis. In other models, the print head moves in the X–Y direction while the print bed moves along the Z-axis.

Rectilinear printers can be divided into two main types: H-Bot and Core X–Y. These types differ in the arrangement of their belts, which move either the printhead or the print bed. In addition, belt printers are specialized in printing extra-long objects (beyond the size of the printer) or in facilitating mass production without interrupting the printing process. In these printers, the print bed is replaced by a belt system and the printhead is fixed at a 45° angle. Special conversion kits are available to convert a standard commercial rectilinear printer into a belt printer.

In addition to rectilinear printers, delta and SCARA (Selective Compliance Assembly Robot Arm) printers are available on the market. These printers differ from rectilinear models in their unique arrangement and movement of the printhead and print bed.

Robotic arms equipped with 3D printheads represent a departure from conventional printers based on the Cartesian system. These arms feature articulated joints similar to those found in human limbs, allowing movement along and around multiple axes. This enhanced range of motion means more degrees of freedom in the 3D printing process, making it easier to produce complex objects. While robotic arm

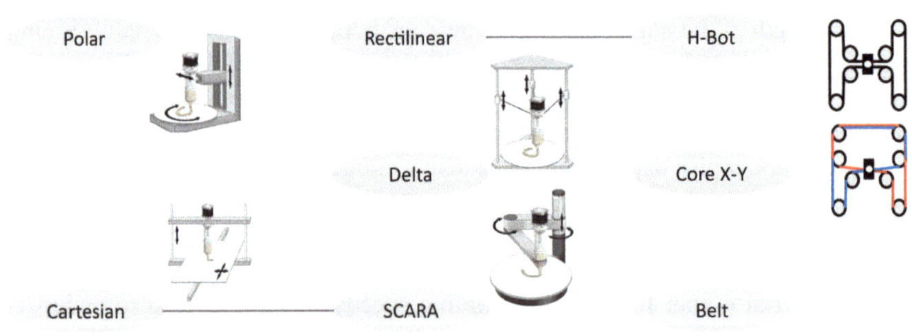

Fig. 20.18 Different printer platforms that can be used for material-extrusion-based additive manufacturing of ceramic-based feedstocks and pastes

printing holds promise for achieving greater flexibility in producing intricate structures, it is important to note that print quality still requires refinement compared to other types of printers. There is also a significant challenge in programming these innovative printers. To date, there are no commercially available slicers specifically designed to program the motion of robotic arms for 3D printing tasks.

20.6.1.4 Multimaterial Printing by AM Thermoplastic Extrusion

One of the key advantages of the MEx-AM thermoplastic ceramic moulding process over other additive manufacturing methods is the ability to print in situ with more than one material to create 3D objects based on multiple ceramic materials. Clemens et al. utilized a commercial printer equipped with two Bowden extruder heads (Sigma BCN3D R17) to fabricate a ZrO2-ZTA 3–3 composite [22]. By adjusting the ceramic filler content within the two feedstocks (e.g. ZrO2- and ZTA-feedstock), they were able to achieve similar shrinkage properties for both materials. However, the process still needs to be optimized to improve mechanical properties. Similarly, Wick-Joliat et al. have demonstrated the use of two auger printheads (e.g. with a customized in-house printer) to produce ceramic objects using two different coloured zirconia materials [29].

At Empa, four printheads are currently in use for various research projects. One of them is a customized in-house printer developed in collaboration with ETH Zurich. Figure 20.19. illustrates the printer, equipped with two direct-drive and two auger extruder heads. The EMPA/ETHZ printer features a tool changer system and an automatic print bed levelling.

20.7 Prospects

Additive manufacturing is rapidly becoming established not only for plastics and metallic materials, and numerous variants are available as alternatives to existing shaping processes like injection moulding, dry pressing, tape casting, etc.

Insofar, in respect of the established processes, it is not incorrect to described additive manufacturing as a “disruptive technology”.

It remains to be seen whether AM will fulfil all the high expectations around the AM hype.

Fig. 20.19 **3D** printers with four MEX-AM printheads, **a** the Tumaker printer has the option of four different printheads, **b** the EMPA/ETHZ printer with four printheads is equipped with a tool changer.

20.8 Conclusions

- In view of the large number of the AM processes and variants existing on the market, careful analysis is required in respect of the product to be manufactured.
- Important criteria for this analysis are the complexity of the object to be printed, geometry, throughput, material, hardening of the "green extrudate" after extrusion, debinding prior to sintering, etc.
- Not every AM process is suitable for printing abrasive and sometimes coarse-grained ceramic materials. Make your CAD data available to the manufacturers and request a printed sample part.
- The possibility of continuously extruding perforated profiles—for instance honeycomb elements—instead of generating the workpiece from discrete points or layer, has been proven as an advantage of AM by extrusion for certain applications.
- Compared to most AM processes, the key features of AM by extrusion are high volumetric throughput rates or short manufacturing times, respectively. That suggests that AM by extrusion is predestined for applications requiring the manufacturing of large-volume objects, like components, walls or houses.

Bibliography

1. W. Kollenberg: Additive Fertigung keramischer Komponenten. Grundlagen und Anwendung, Essen 2020
2. A. Ghazanfari, W. Li, M. C. Leu, G. E. Hilmas: "A novel extrusion-based additive manufacturing process for ceramic parts" In: Proceedings of the 27th Annual International Solid Freeform Fabrication Symposium—An Additive Manufacturing Conference, (As pdf available via Internet.)
3. U. Scheithauer, E. Schwarzer, T. Moritz, A. Michaelis: "Heat Exchanger: Opportunities and Limits of the Lithography-Based Ceramic Manufacturing (LCM)." In: Journal of Materials Engineering and Performance, Volume 26,1, 2017.
4. A. Zocca, P. Colombo, C. M. Gomes, J. Günster: "Additive Manufacturing of Ceramics: Issues, Potentialities, and Opportunities". In: Journal of the American Ceramic Society, Volume 98, Issue 7.
5. Zhangwei Chen, Ziyong Li, Junjie Li, Chengbo Liu, Changshi Lao, Yuelong Fu, Changyong Liu, Yang Li, Pei Wang, Yi He: "3D printing of ceramics: A review". In: Journal of the European Ceramic Society, Volume 39, Issue 4.
6. J. Abel, U. Scheithauer, T. Janics, S. Hampel, S. Cano, A. Müller-Köhn, A. Günther, Ch. Kukla, T. Moritz (2019): "Fused Filament Fabrication (FFF) of Metal-Ceramic Components", January 11, 2019, https://www.jove.com/de/v/57693/fused-filament-fabrication-fff-of-metal-ceramic-components, February 26, 2024.
7. F. Händle, F. Laenger, J. Laenger „Ermittlung der Pressdrücke bei der Extrusion keramischer, Pressmassen mit Hilfe der Benbow-Bridgwater-Gleichung unter Verwendung des Capillar-Checks", In: cfi/Ber. DKG 92 (2015) No. 9
8. F. Händle, M. Simon: "Additive Manufacturing by Extrusion in Ceramics" In:cfi.Ber.DKG97(2020) No 3–4
9. Carneau, Paul; Mesnil, Romain; Baverel, Olivier; Roussel, Nicolas: Layer pressing in concrete extrusion-based 3D-printing: Experiments and analysis. In: Cement and Concrete Research Bd. 155 (2022), Nr. 1, S. 1–14
10. Mechtcherine, Viktor ; Nerella, Venkatesh Naidu ; Will, Frank ; Näther, Mathias ; Otto, Jens ; Krause, Martin: Large-scale digital concrete construction—CONPrint3D concept for on-site, monolithic 3D-printing. In: Automation in Construction Bd. 107 (2019), Nr. 1, S. 1–16
11. Reinold, Janis ; Nerella, Venkatesh Naidu ; Mechtcherine, Viktor ; Meschke, Günther: Extrusion Process Simulation and Layer Shape Prediction during 3DConcrete- Printing Using The Particle Finite Element Method. In: Preprints (2020), S. 1–25.— https://doi.org/10.20944/preprints202007.0715.v1
12. Clemens, F. J. *et al.* (2020). FDM/FFF an Alternative to CIM Manufacturing of Prototype and Small Quantities of Ceramic Part. Ceram Appl, 8, 27-31.
13. Standard, F2792. 2012. standard terminology for additive manufacturing technologies, ASTM F2792–10e1, 2012.
14. McNulty, T. F. *et al.* (1998). Development of a binder formulation for fused deposition of ceramics. Rapid Prototyping Journal, 4(4), 144-150.
15. Allahverdi, M. *et al.* (2001). Processing of advanced electroceramic components by fused deposition technique. Journal of the European Ceramic Society, 21(10-11), 1485-1490.
16. T Crump, S. (1989). Method for Creating Three-Dimensional Objects. US Patent US5121329A.
17. Crump, S. (1992). US5121329 A-Apparatus and method for creating three-dimensional objects.
18. Greul, M. *et al.* (1995). Rapid prototyping of functional metallic parts. Computers in Industry, 28(1), 23-28.
19. Jafari, M. *et al.* (2000). A novel system for fused deposition of advanced multiple ceramics. Rapid Prototyping Journal, 6(3), 161-175.
20. Atisivan, R. *et al.* (2001). Porous mullite preforms via fused deposition. Journal of the American Ceramic Society, 84(1), 221-223.

21. Scheithauer, U. *et al.* (2015). Thermoplastic 3D printing—an additive manufacturing method for producing dense ceramics. International journal of applied ceramic technology, 12(1), 26-31.
22. Clemens, F. *et al.* (2021). Debinding and sintering of dense ceramic structures made with fused deposition modeling. In Industrializing Additive Manufacturing: Proceedings of AMPA2020 (pp. 293–303). Springer International Publishing.
23. Bach, M. *et al.* (2018). Additive manufacturing of piezoelectric 3–3 composite structures. In Industrializing Additive Manufacturing-Proceedings of Additive Manufacturing in Products and Applications-AMPA2017 1 (pp. 93–103). Springer International Publishing.
24. Gorjan, L. *et al.* (2018). Ethylene vinyl acetate as a binder for additive manufacturing of tricalcium phosphate bio-ceramics. Ceramics International, 44(13), 15817-15823.
25. Clemens, F. *et al.* (2023). Material extrusion additive manufacturing of advanced ceramics: Towards the production of large components. Journal of the European Ceramic Society, 43(7), 2752-2760.
26. Sarraf, F. *et al.* (2023). EVA-PVA binder system for polymer derived mullite made by material extrusion based additive manufacturing. Journal of the European Ceramic Society, 43(2), 530-541.
27. Bellini, A., *et al.* (2005). New developments in fused deposition modeling of ceramics. rapid prototyping journal, 11(4), 214–220.
28. Hadian, A. *et al.* (2022). Material extrusion additive manufacturing of zirconia parts using powder injection molding feedstock compositions. Additive Manufacturing, 102966.
29. Wick-Joliat, R., *et al.* (2023). Multi-material ceramic material extrusion 3D printing with granulated injection molding feedstocks. Ceramics International, 49(4), 6361-6367.

Chapter 21
Simulation in Ceramic Extrusion

All models are wrong, but some models are useful
—George P. E. Box.

21.1 Introduction

This chapter will deal with the topic of CFD simulations (computational fluid dynamics) in the context of ceramic extrusion. As extrusion processes are not transparent most of the time, it is difficult to gauge what is really going on inside your extruder. CFD simulations may help to give you an insight into your process and increase understanding of the dominant flow phenomena. This makes optimization of the process much easier and faster. In addition, it enables testing of modifications to the extrusion set-up without the need for further experiments in the laboratory, which can save time and potentially lower development costs. Therefore, on the one hand, CFD simulations can be a powerful tool in research and development as well as in the optimization of existing equipment. On the other hand, using CFD techniques and interpreting the results correctly can be a challenging task that requires some prior knowledge. This chapter will give some advice on how to use CFD techniques wisely and which results can be reasonably expected. The chapter is structured as follows:

The basic working principle of a general CFD simulation will be roughly discussed in Sect. 2 of this chapter. Section 3 will deal with boundary conditions and some ways to determine necessary model parameters. Section 4 lists typical mistakes made in CFD simulations that can be avoided by taking some simple precautions. Section 5 finally shows some real-world examples where CFD simulations have helped to improve process understanding and process performance, and Sect. 6 will summarize the overall content.

F. Händle, *The Art of Ceramic Extrusion*, https://doi.org/10.1007/978-3-031-63471-0_21

21.2 The Principle of CFD Simulations

The basic principle of every CFD code is to start with physical principles and transform them into calculable equations. These principles are, namely, the conservation of momentum in all three directions of space, the conservation of mass and the conservation of energy. Solving these equations, at least approximately, gives access to the velocity profile, pressure field and temperature distribution at almost every location inside your computational domain. A very detailed and comprehensible derivation of the mathematical equations from the universal conservation laws can be found in [1]. By assuming the flow to be incompressible (ρ = const.) and using Fourier's law to model diffusive energy transport, the resulting set of differential equations for describing the fluid reads as follows:

$$\frac{\partial}{\partial t}\vec{u} + \nabla(\vec{u} \otimes \vec{u}) = -\frac{1}{\rho}\nabla p + \frac{1}{\rho}\nabla\tau^{\circ} + \vec{f}$$

$$\nabla\vec{u} = 0$$

$$\frac{\partial}{\partial t}e + \nabla\left(e \cdot \vec{u}\right) = \dot{q} + \frac{1}{\rho}\nabla(\mathrm{k} \cdot \nabla \mathrm{T}) + \frac{1}{\rho}\tau : \nabla\vec{u}$$

In this set of equations $\vec{u}$ represents the velocity vector [m/s], p is the pressure [Pa], ρ is the density [kg/m^3], τ is the stress tensor [Pa] inside the fluid and $\vec{f}$ is an external body force [m/s^2], such as gravity for instance. The energy equation sets the inner energy e [J/kg] in relation to the volumetric external heating source $\dot{q}$ [J/kg/s], the thermal conductivity k [W/m/K], the Temperature T [K] and other quantities already known from the momentum balance. In this notation the $\otimes$stands for the dyadic product of two vectors to build a tensor and the: operator in the energy balance represents the tensor scalar product, also known as double inner product (e.g. $\tau : \nabla\vec{u} = \tau_{xx} \cdot \frac{\partial u}{\partial x} + \tau_{yx} \cdot \frac{\partial u}{\partial y} + \cdots + \tau_{yz} \cdot \frac{\partial w}{\partial y} + \tau_{zz} \cdot \frac{\partial w}{\partial z}$ with u, v and w being the velocity components of $\vec{u}$). To make these equations solvable, the stress tensor τ must be linked to material properties and the other solution variables (e.g. $\vec{u}$) in some way while the inner energy e must be somehow represented as a function of temperature T. Both closures will be addressed separately in this section.

Finding a good representation for the stress tensor τ can be difficult. For very simple Newtonian fluids, the formula for τ can be written as

$$\tau = \eta \cdot 2 \cdot D$$

with the Newtonian viscosity η [Pa*s] and the deformation tensor $D = \frac{1}{2}(\nabla\vec{u} + \nabla\vec{u}^T)$. . Unfortunately, most fluids do not show Newtonian behaviour. Complex models have been developed to describe the rheology of different media. The Ostwald- and Carreau-models are common choices to represent the rheology of polymer melts [11], while the Bingham- and Herschel–Bulkley-models are used to describe ceramic materials, because they are able to represent the mostly observed yield stress τ_0. . The yield stress of a fluid causes it to behave like a solid if the applied stress is not sufficient. Examples of yield stress fluids besides most ceramic materials are bitumen [11], toothpaste, printing ink, ketchup or margarine [6]. You have to apply a minimum amount of stress (called τ_0) before these even start to shear. Above this minimum stress level, they show additional stress against shearing that is related to the shear rate $\dot{\gamma}$ and a viscosity-like value η_{Bi}. The shear rate itself is derived from the deformation tensor D:

$$\dot{\gamma} = \sqrt{2} \cdot \sqrt{D : D}$$
$$= \sqrt{2} \cdot \sqrt{\left(\frac{du}{dx}\right)^2 + \frac{1}{4}\left(\frac{du}{dy} + \frac{dv}{dx}\right)^2 + \ldots + \frac{1}{4}\left(\frac{dv}{dz} + \frac{dw}{dy}\right)^2 + \left(\frac{dw}{dz}\right)^2}$$

It should be noted that some literature sources cancel out the factors 2 and ½ in the definition of τ and D, which leads to a similar (but different) definition $\dot{\gamma} = \frac{1}{\sqrt{2}} \cdot \sqrt{D : D}$. We advise taking the written-out definition on the right-hand side, which is true and independent of definition of D. A better understanding of the different rheology models can be achieved in two dimensions, where the rheology equations can be written as

$$\tau = \eta \cdot \dot{\gamma}$$

for Newtonian fluids and

$$\tau = \tau_0 + \eta_{Bi} \cdot \dot{\gamma}^n$$

for Herschel–Bulkley fluids with n as an additional model parameter to describe shear-thinning ($n < 1$) or shear-thickening ($n > 1$) behaviour. A value of $n = 1$ gives the original Bingham rheology model. If the stress applied to the fluid is smaller than the yield stress τ_0, the fluid does not flow and $\dot{\gamma}$ has to be zero. This makes a numerical solution difficult as you cannot simply write an if–then-condition into an implicit algebraic equation system. To get a good numerical performance, you want to solve an equation system similar to the case of a Newtonian fluid. This structure is achieved by rewriting the Herschel–Bulkley model into a yield-stress-free rheology but with spatially varying viscosity. The numerical code solves the same three-dimensional equation system as in case of a Newtonian fluid ($\tau = \eta \cdot 2 \cdot D$) but with a non-constant viscosity field that accounts for the yield stress. This approach makes the equations easier to solve for the computer but introduces a small modelling error in places where the fluid should remain unyielded, because the effective viscosity represents the resistance to deformation but cannot avoid it completely. If a fluid element should remain unyielded, the viscosity at this place has to be infinitely large, which causes numerical problems. To get back the original definition of a Herschel–Bulkley fluid, the effective viscosity would have to be

$$\eta = \frac{\tau_0 + \eta_{Bi} \cdot \dot{\gamma}^n}{\dot{\gamma}}$$

This viscosity definition will fail numerically when shear rate $\dot{\gamma} \to 0$ because the viscosity would tend to infinity. This problem is solved by allowing a certain amount of shearing even in places where the medium should show no shearing. For this purpose, the viscosity equation is modified by restricting the maximum reachable viscosity value to a certain limit η_{max} that can be handled by the CFD code.

$$\eta = \min\left(\frac{\tau_0 + \eta_{Bi} \cdot \dot{\gamma}^n}{\dot{\gamma}}; \eta_{max}\right)$$

The maximum value η_{max} typically orients to the value of η_{Bi}. If $\eta_{max} = 100 \cdot \eta_{Bi}$ you can roughly expect that the erroneous local velocity gradient in an otherwise unyielded region is 100 times smaller than the velocity gradient at the nearest wall of your system. Checking shear rate plots of your results will give you a good impression here because they are directly related to velocity gradients. Shear rates that are 100 times smaller than the maximum shear rate in the vicinity might be artefacts of your choice $\eta_{max} = 100 \cdot \eta_{Bi}$ and should be zero instead. By choosing $\eta_{max} = 1000 \cdot \eta_{Bi}$ you can roughly expect the erroneous velocity gradients and artificial shear rates to be 1000 times smaller instead. This modelling error has to be known when CFD results are interpreted. A different approach to rewrite the viscosity function can be found in the work of Papanastasiou, who limits the effect of τ_0 in areas of small shear rates instead of limiting the final viscosity value to a maximum level [12].

$$\eta = \frac{\tau_0 \cdot (1 - \exp(-m\dot{\gamma})) + \eta_{Bi} \cdot \dot{\gamma}^n}{\dot{\gamma}}$$

This equation needs another parameter m that should be chosen as large as possible to get as close as possible to the original Herschel–Bulkley model. Values of $m = 50$ or $m = 100$ are reported in the work of Papanastasiou [12], but the best value depends on the performance of the numerical code and the accuracy requirements of your results. The best way to estimate the introduced error is to run the simulation twice with different values for m and compare the results. If the relevant results change significantly, you have to increase m even further until the changes are negligible. To stabilize the numerical behaviour, the shear rate in the denominator is often limited to a minimum value in the order of magnitude of machine precision (typically $\sim 10^{-16}$) to avoid dividing by zero during iteration of while iterating the equation system. Further discussion about rheological modelling and implementation in CFD codes can be found in the review part of [4]. An example from literature that uses this approach of effective viscosity to model Herschel–Bulkley fluids in CFD simulation can be found in [9].

Following the discussion of the closure for the stress tensor τ, the definition of the inner energy e should be addressed. An easy representation can be given by introducing the specific heat capacity c_V [J/kg/K], which must be measured in some way and generally depends on temperature as well. Then an expression for e can be given that is valid at least for small temperature changes by

$$e = e_0(T_0) + c_V(T_0) \cdot (T - T_0)$$

with a reference temperature T_0 at which c_V and e_o were measured. More complex equations are possible for e like higher-order polynomials or other functions, but that is a topic for literature about thermodynamics and is only relevant if the observed temperature range is larger than a few degrees K, so that c_V changes significantly. Dependence on pressure might be possible for some systems, but that would make the assumption of incompressible flow questionable and will not be addressed here.

21.3 Determining Model Parameters and Boundary Conditions

After closing of the equation system to make it solvable, the correct boundary conditions must be assigned to get the correct results. Relevant to mention at this point are the walls and the inlet section of your computational domain, while outlet conditions are quite clear. At the inlet, the most difficult issue is assigning a correct value for the inflow. Two approaches exist herein. In the first approach you can assign a known flow rate or velocity to the inlet. This value can be easily obtained if you are using a piston press. The pressure field will be calculated by the CFD solver. The other approach assigns a pressure value at the inlet. The CFD code then calculates the flow rate that will fulfil the pressure drop between inlet and outlet. This approach is typically used for screw extruders when you have measured pressure values right in front of the entrance of the extrusion die. In both approaches you need to assign one variable (pressure or velocity) while the other is calculated by the CFD code. You cannot assign both. The energy equation needs a certain inlet temperature that has to be known and will be converted to the corresponding value for the inner energy e by the code. More interesting than the inlet are the walls of your extrusion die. Regarding the energy equation, you have to decide if energy transfer to the walls is relevant or not. If it is not important, the adiabatic boundary condition can be applied. Otherwise, if you know from practice that you need heating or cooling during extrusion, it definitely is relevant. In this case you would need to assign a heat flux or a fixed temperature value to the wall. As the thermal conductivity of metals is much larger than the thermal conductivity of ceramic materials, it is common to set a fixed temperature value to the metallic wall that is equal to the environment or the used cooling/heating utility. The more accurate way would be to include the wall material as a solid into the computational domain and solve the energy equation inside the wall as well. This might be necessary if the thermal conductivity of your extrusion material is in the same order of magnitude as the thermal conductivity of your wall. Beside the thermal boundary condition, the most difficult choice lies in the boundary condition for the velocity. As many ceramic materials tend to slip at the wall, this behaviour must be formulated in a mathematical equation to tell the code how to deal with it. One of the first and probably most famous attempts to determine wall slip can be found in [10]. Mooney describes a certain way to measure the magnitude of wall slip without knowing the underlying rheology model. By pressing the fluid through circular capillaries with different radii, while keeping the wall shear stress τ_{wall} constant in some way over all used capillaries, you can identify the slip velocity u_{slip} at this certain value of wall shear stress. This procedure can be repeated at different wall shear stresses to scan the amount of wall slip over the complete range of relevant values of τ_{wall}. A demonstration of this kind of measurement and data processing procedure in principle can be found in [5] and [7] respectively. This gives a certain data set to describe at which wall shear stress which amount of wall slip u_{slip} occurs. Nevertheless, the CFD code needs a continuous function as input to describe the wall slip ($u_{slip} = f(\tau_{wall})$) that is in reasonable agreement with measured wall slip values. This function can have any arbitrary form. Most of the time, a linear correlation is found or assumed like $u_{slip} = a_0 \cdot (\tau_{wall} - \tau_{slip})$ with two coefficients a_0 and τ_{slip} that have to be determined from the data set by regression. More complex functions are possible like $u_{slip} = a_0 \cdot (\tau_{wall} - \tau_{slip})^s$ or anything else that is physically arguable. A more general description of wall slip can be found in the literature review part of [3]. Stating that wall slip and rheology obey certain mathematical equations, which is necessary to define a valid CFD model anyway, offers a second possibility to determine all necessary model coefficients. For most mathematical descriptions, the resulting isothermal momentum balance for axial flow in a capillary including the wall slip condition can be solved analytically, as demonstrated in [3]. This gives a clear theoretical relation between pressure drop and throughput in a capillary depending on the assumed model coefficients a_0, τ_{slip}, s, τ_0, η_{Bi} and n. If measurement and other experimental errors are neglected, there should be a set of model coefficients such that the measured pressure drop exactly fits to the theoretical pressure drop that is predicted by the analytical relation for every experimental throughput. A regression model can be formulated to find the best set of model coefficients (a_0, τ_{slip}, s, τ_0, η_{Bi} and n) to minimize the discrepancy between measured pressure drop and predicted pressure drop for all measured data points. This coefficient set gives the most probable description of the extruded medium in that shear rate regime assuming that rheology and wall slip can be described by the chosen equations. It should be noted that the regression can also be formulated the other way around, assuming the pressure drop to be given and minimizing the discrepancy between observed and predicted throughput. This might be easier depending on the form of the analytically found prediction formula. To make the regression result usable over a big range of wall shear stresses and shear rate regimes, it is important to investigate a wide spectrum of radii and throughputs during the experimental measurements.

21.4 Error Estimation

All this described effort is necessary to formulate a valid and really accurate CFD model. The quality of the CFD results will depend heavily on the quality of the model input. Performing some rough estimations of material data will result in an uncertainty of the simulation results as well. Sometimes this principle is snidely called GIGO, which stands for "garbage in, garbage out", but could also be interpreted as "good in, good out". Depending on the aim or question addressed by the CFD simulation, a vague result might be sufficient, but also has to be interpreted in that fashion, so don't trust in the fifteenth digit of your CFD results if the input data are only accurate to two digits. For example, if the specific heat capacity has an error of about 10%, this will lead to an error in evaluation of temperature rises in approximately the same order of magnitude. Similar estimations can be made for wall slip coefficients or viscosity parameters influencing the predicted pressure drop. A CFD code translates a mathematical description of your system into better understandable data (e.g. pressure, velocity and temperature fields). It does not do more, but does not do less than this either, which leads to a general discussion about possible errors of CFD simulations. To reiterate, they can be a very powerful tool if used wisely. Therefore, the most common mistakes should be addressed here to rule these out a priori. According to Ferziger and Peric, there are four main types of errors associated with simulation results [2].

21.4.1 Programming Error

The first error is called programming error, which can come from wrong implementation of the underlying maths. In the case of commercial CFD codes, it is out of reach for the end-user to check that. Nevertheless, if you have implemented a new wall slip condition in your model based on your own written code, it is advisable to validate this implementation by performing a simulation where you already know the result. One of the analytical solutions for capillary flow, taken from [3], might be useful here.

21.4.2 Iteration Error

The second common error is called iteration error. CFD codes transform differential equations into linear algebraic equation systems which are too large to solve with the methods taught in school like Gaussian Elimination or Cramer's Rule. Iterative methods are used instead to solve the equation system approximately. These approximations get closer and closer to the true solution as more and more iterations are performed, but that is time-consuming. A tolerance must be given to the CFD solver stipulating to which accuracy the equation system has to be solved. If this tolerance is too loose, this might influence the results. This type of error can be ruled out by performing additional simulations with tighter tolerances until the results no longer change. A typically good starting point for this tolerance is to force a reduction of residuals by at least four orders of magnitude and compare with the results with a reduction of five orders of magnitude.

21.4.3 Discretization Error

The third error source in CFD simulations is called the discretization error. It is directly linked to the underlying mesh. The mesh dictates in some sense the spatial sampling rate of your solution. You will get information about pressure, velocity and temperature in every cell centre of your CFD mesh, but if some interesting phenomena happen in the space between two cell centres, this will not be contained in your results. A good start to estimate the necessary amount of grid cells is to imagine how the real solution is constructed from piecewise elements within each cell. In most finite volume methods, the profile of each variable within each cell or between two cell centres is approximated by linear functions. You can imagine that you need a certain amount of straight-line segments to build a curved parabolic flow profile. Finite element methods sometimes use quadratic functions instead to approximate the profile within each cell. Then a single cell in height might be sufficient to get 100-% accurate results for Newtonian laminar flow between two plates. In the case of solution profiles that are not perfectly parabolic, even high-order finite element methods need more cells, but typically fewer than finite volume methods. Therefore, it is useful to know the basic approximation technique of your CFD code to estimate the necessary number of elements. The easiest way to validate your choice is to perform the CFD simulation on a fine and a coarse grid. Refine the grid further until the results do not change any more or until the results do not change more than your acceptable error tolerance for the results. If you refine the mesh by a factor of two in every spatial dimension, you can expect that the discretization error will reduce by a factor of 2 – 8 based on the discretization technique you have used. It is common to choose different mesh sizes in different regions of your computational domain to decrease discretization error locally where needed. The same principles hold for the temporal discretization error if you run a time-dependent simulation. Almost every time the highly viscous flow in an extrusion die will be steady state so that temporal discretization errors will not be a factor, but they should be kept in mind for transient simulations and can be investigated by running a reference simulation with smaller time step size.

21.4.4 Modelling Error

The last typical error source in CFD results comes from wrong model definition. CFD codes produce data based on mathematical models. If the model itself does not describe the system behaviour well, the results will not do so either. Therefore, it is so important to use the correct boundary conditions for your system and to choose appropriate equations to describe wall slip and rheology. Unfortunately, there is no easy way to estimate the error size of wrong modelling. But it can be argued that errors should be small providing certain requirements are fulfilled. Some of these are listed below

- Regarding wall slip, the measurements to determine model coefficients should be made at the same wall shear stress level as expected to exist in the final extrusion die.
- Rheology measurements should be taken in the same regime of true shear rate as expected to exist in the final geometry. Measuring in capillaries with similar diameter and with similar throughputs/pressure drops as in your target case does not guarantee the same shear rate regimes, but helps to hit the correct order of magnitude for $\dot{\gamma}$.
- If a Herschel–Bulkley- or Bingham model was used for rheology, it should be checked if the values of m or η_{max} were sufficiently high to mimic the real system behaviour. If there is any doubt, a second simulation with higher values should be run for comparison.
- Certain parameters like viscosity are temperature-dependent. They should be implemented that way or it is necessary to verify in another way that the changes are negligible.
- All equations mentioned in this chapter are formulated based on continuum mechanics. This assumes a homogeneous mixture of the extrusion medium. Any large grains or undissolved agglomerates will make the basic formulation of momentum an energy balance fail.

This is not a conclusive list but helps to rule out the most common modelling errors.

21.5 Application of Simulation Methodology

The simulation of the extrusion of ceramic bodies is proving to be crucial for the success of the modern ceramics industry. Both from an economic point of view, through the realization of cost savings and quality improvements, and due to current market developments, that require adaptation to new requirements and material innovations, the importance of simulation is becoming increasingly obvious. With rising energy prices, resource scarcity and skills shortages, only companies that invest in this technology will be able to meet the increasing demand for high-quality, customized ceramic products and stand out in a competitive market. The following are therefore some examples that would not have been possible without the use of simulation methodology adapted to ceramic masses.

21.5.1 Ceramic Masses in Battery Systems

In the rapidly growing electric mobility industry, batteries are crucial to optimizing the range, performance and longevity of electric vehicles. To meet these requirements, ceramic masses are playing an increasingly important role in modern battery systems for automobiles. This requires extremely thin ceramic foils, which are then used in the battery. Classically, these are calendered by means of a rolling mill, which can, however, lead to contamination and operator-dependent fluctuations in quality. To avoid this, wide slot dies are to be used in the future, as they are also known from plastic film extrusion. With the help of the simulation methodology shown above, these can be designed geometrically and process-technically in such a way that a homogeneous discharge takes place and a uniform film can be extruded (Figs. 21.1 and 21.2).

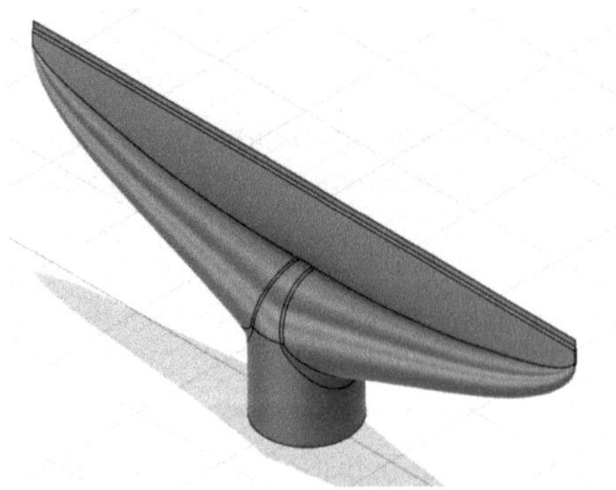

Fig. 21.1 Flow channel (negative) of a coat-hanger distributor for film extrusion

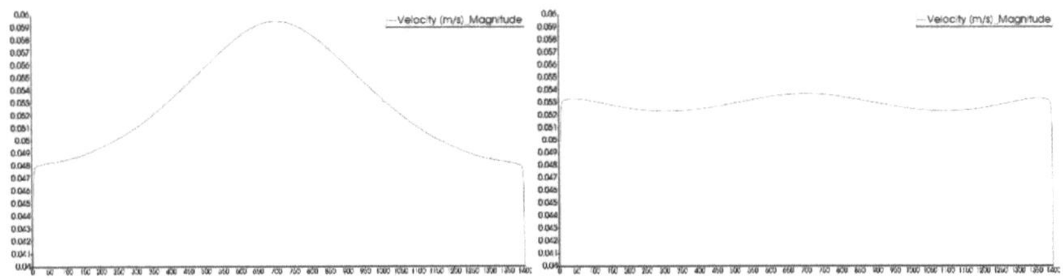

Fig. 21.2 Outlet profile with classical wall adhesion (left) and revised simulation methodology (right). The velocity in [m/s] on the y-axis is shown above the exit width in [mm] on the x-axis

The fluid area of the corresponding coat-hanger distributor is shown below.

By applying classical simulation methods with wall adhesion as well as the adapted simulation methodology, clearly different simulation results are obtained:

The figures show the exit profile in the form of the velocity at the exit of the die. The goal must be a velocity distribution that is as homogeneous as possible, so that a film with a homogeneous cross-section leaves the mould. The classical simulation methodology makes it appear that the die is extremely inhomogeneous. The revised simulation methodology, on the other hand, shows an almost flat exit profile, which has been confirmed in investigations. This shows that the application of a simulation model adapted to ceramic extrusion is critical for the design of extrusion tools in the ceramic industry. Failure to observe this leads to incorrect design and iteration loops, which should be avoided by the simulations.

21.5.2 Extrusion of Profile Geometries

Another example in which simulation is used for design and iteration loop avoidance are run-in processes of profile dies—i.e. dies that convert the circular ceramic stream provided by the extruder into a specific contour, possibly with hollow chambers and different wall thicknesses. In design without simulation, several iteration loops are usually necessary on the already built tool in order to achieve the desired product quality. With prior simulative optimization, these iteration loops can be performed virtually, saving valuable resources in the form of time, money and material. Again, however, it is important to work with the correct simulation model. Again, the exit profile of an extrusion die is shown below (Fig. 21.3).

The calculation using the classical wall-adherent model showed that the inner ribs are flowed through the least. A correspondingly optimized model would have resulted in an enlargement of the cut-outs for the ribs, so that the material flow is increased. The adapted model, however, achieved exactly the opposite result. Here, the flow showed a maximum in the middle, and further opening of the ribs would have increased the inhomogeneity. Through a simulative optimization of the geometry, the exit velocity could be further adjusted.

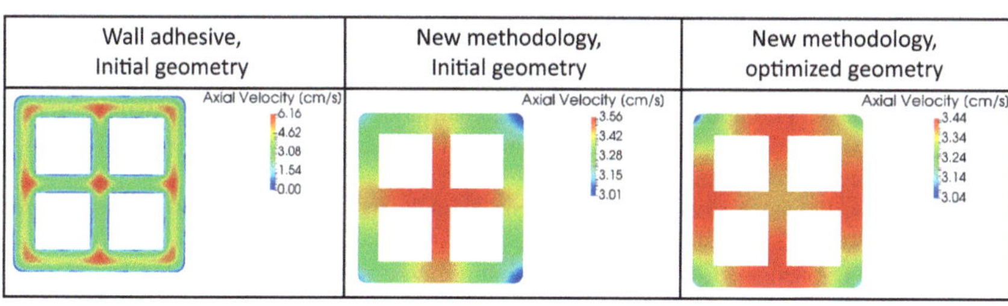

Fig. 21.3 Simulative exit velocity of a profile tool in ceramic extrusion

21.5.3 Calculation of Extrusion Screws

Also, in the calculation of screws, which homogenize the ceramic mass and apply the necessary pressure to pass through the die, the correct modelling of the material behaviour is of utmost importance. Especially in the calculation of the possible pressure build-up of an extrusion screw, the rheological behaviour is very important. Thus, in the classical calculation with no-slip rheological models, pressure build-ups are greatly overestimated due to the very good conveying effect. In the worst case, a screw is designed which in reality cannot apply this pressure at all and thus cannot provide the desired mass flow for the process. An example of simulative investigation inside an extrusion screw with self-written code can be found in [8].

21.6 Conclusion

In summary, CFD simulations are potent tools to gain insight into your extrusion process and increase your understanding with regard to possible improvements. With the use of modern software packages, they can be done by almost anyone. The challenging part is to build a good mathematical model description of your system by choosing correct boundary conditions and reasonable rheological models with corresponding parameters. Translating the mathematical model into humanly readable data like pictures and animations is a nearly completely solved task and will be performed by the CFD code. Feeding the code with the correct settings is the human task that can easily become the dominant error source. Hopefully, this chapter has given sufficient insight into and understanding of the concept of CFD codes so you can give your attention to the most important aspects when applying simulations in practice.

Bibliography

1. J. Anderson, Computational Fluid Dynamics: The Basics with Applications (McGraw-Hill Education (1995)), ISBN 0–07–001685–2
2. J.H. Ferziger, M. Peric, Computational Methods for Fluid Dynamics (Springer-Verlag, Berlin, (2002)), ISBN 3–540–42074–6
3. Y. Damianou, M. Philippou, G. Kaoullas, G.C. Georgiou, Cessation of viscoplastic Poiseuille flow with wall slip (Journal of Non-Newtonian Fluid Mechanics, Volume 203, p24–37, (2014)), https://doi.org/10.1016/j.jnnfm.2013.10.004
4. E. Mitsoulis, Flows of viscoplastic materials: models and computation (Rheology reviews 2007, p135–178, (2007))
5. F. Haendle, F. Laenger, J. Laenger, Determining the Forming Pressures in the Extrusion of Ceramic Bodies with the Help of the Benbow-Bridgwater Equation Using the Capillar Check (ceramic forum international, No. 10–11, (2015))
6. J.C. Misra, S.D. Adhikary, Flow of a Bingham fluid in a porous bed under the action of a magnetic field: Application to magneto-hemorheology (Engineering Science and Technology,

an International Journal, Volume 20, Issue 3, June 2017, p973–981, (2017)), https://doi.org/10.1016/j.jestch.2016.11.008
7. K.-F. Laenger, F. Laenger, K. Geiger, Wall Slip of Ceramic Extrusion Bodies, Part 2 (ceramic forum international, No. 4–5, (2016))
8. J. Laenger, F. Laenger, Wall Slippage of Ceramic Bodies, Part 3 - The New ESM 4.0 from cpe (ceramic forum international, No. 3, (2018))
9. S. Lovato, G. H. Keetels, S.L. Toxopeus, J.W. Settels, An eddy-viscosity model for turbulent flows of Herschel–Bulkley fluids (Journal of Non-Newtonian Fluid Mechanics, Volume 301, 104729, (2022)), https://doi.org/10.1016/j.jnnfm.2021.104729
10. M. Mooney, Explicit formulas for slip and fluidity (J. Rheol. 2, 210–222, (1931)), https://doi.org/10.1122/1.2116364
11. M. Pahl, W. Gleissle, H.-M. Laun, Praktische Rheologie der Kunststoffe und Elastomere (VDI-Verlag GmbH, Düsseldorf, (1995)), ISBN 3–18–234192–8
12. T.C. Papanastasiou, Flows of Materials with Yield (J. Rheol. 31, 385–404, (1987)), https://doi.org/10.1122/1.549926

Chapter 22
A Look-Ahead

We tend to overestimate the effect of a technology in the short run and underestimate the effect in the long run.
—Roy Amara

Abstract The future of ceramic extrusion is wide open and offers great chances and challenges. Three basic strategies are outlined how to tackle the future as well as three gaps/blind spots in our know-how, theories and technology. To follow the elements of the "Smart Extrusion Concept" (SEC) might be a useful option to master the future of ceramic extrusion.

Keywords Strategies for extrusion · Gaps and blind spots · Know-how · Wear protection · Sensors · Simulation · Smart extrusion concept

22.1 Predictions, Strategies and Wishes

We have to realize that any kind of prediction is problematic and that coincidence plays a much greater role, more than we admit, both in life and also in terms of technology. Despite this, there exists a whole industry of the so-called experts and gurus of different origin, who all make a living from prognosticating non-stop. Most of their predictions end up in the rubbish basket. In fact, we find ourselves in a dilemma because we are constantly forced to make forecasts for the next day, or the next week or the next year, in order to prepare ourselves for possible eventualities. The dilemma is aggravated by rapid technological progress, which I do not need to go into here: big data, deep learning, augmented reality, Internet of things, nanotechnology, additive manufacturing, robotics, artificial intelligence, etc. What are the possible scenarios for which we could or should prepare ourselves in the extrusion of ceramic bodies? Could it be possible for ceramic honeycombs and other profiles to be produced by additive manufacturing, as believed by euphoric experts? Are there other so-called disruptive technologies in sight? As a producer of ceramic extrudates or designer of extrusion systems, we have to think about the next steps there may be within the sector. In the end, it comes down to the allocation of resources in the absence of

F. Händle, *The Art of Ceramic Extrusion*, https://doi.org/10.1007/978-3-031-63471-0_22

complete information and under time pressure. Basically, there are three strategies available to us:

- Strategy 1 "enhancement": constantly and jointly with pertinent competent partners review possible enhancement potential in terms of efficiency and/or product quality that could be implemented in existing systems. Strong process and customer orientation are important pillars of this strategy. The "Toyota Production System" with its components Kaizen, Jidoka, PDCA, SMED, etc. is typical example of this strategy.
- Strategy 2 "wish list": within the second strategy, we develop scenarios by compiling a catalogue of features which describe a perfect extrusion system; this should be as accurate as possible and should also contain features which are not available at the time of compilation. Part of this exercise is also to name the "blind spots" in our knowledge of ceramic extrusion. This strategy can be of a great instrumental use as much as the items on the "wish list" can be worked through one by one. Those which cannot be realized at the time should be updated and revised over and over again. Let me quote a metaphor used by a most successful Italian entrepreneur and business friend: "I am like a fox sitting on a riverbank, waiting for the very thing I want to just drift by".
- Strategy 3 "muddling through": by far the most popular strategy is known by the name "management by muddling through". It is not necessary to describe this in greater detail; it might well be successful in the short run, but definitely not in the long run.

As we are quite familiar with the "muddling through" strategy and can well imagine what is to be understood by continuous enhancement in extrusion, I shall, in the following, focus on the "wish list" strategy, confining myself to only three wishes/gaps (Fig. 22.1).

22.2 Gaps and Blind Spots

22.2.1 The Know-How Gap

In recent years, we have learnt a great deal regarding the extrusion of ceramic bodies. This also explains the progress achieved in the extrusion of extremely fragile profiles such as honeycombs and difficult ceramic bodies. However, now as before, we have no thorough knowledge of the combined rheological, tribological, chemical, mechanical and thermal processes which take place on a microscopic scale and which influence each other. Strictly speaking, the extrusion of ceramic bodies is from a strict scientific viewpoint still a "black box". This means that what we do is to extrapolate from the results of the processes which have taken place, then deduce optimization

Fig. 22.1 Look-ahead into the vague future of extrusion

measures and use the adjustment screws available to us relating to the body, the preparation process, the die, the geometry of the auger, etc. As explained in Chap. 4, there do exist theoretical approaches but there is no "uniform extrusion theory" available.

The most promising route from the "Art of Extrusion" to the "Science of Extrusion" is probably the simulation approach. That is, we simulate the flow of the ceramic body respectively the ceramic particles through the different zones of the extruder right up to the die. Detect the speed and temperature gradients in the different zones to find out where segregation and dewatering happens, and "dead corners" or laminations are generated. Suitable approaches and algorithms/simulation models are already in existence, but in the run-up to such simulations, it is however always necessary to precisely determine the rheological properties and corridors of the ceramic bodies to be processed.

22.2.2 *The Wear Protection Gap*

Progress has been achieved over recent years in terms of wear protection as reported in Chap. 15. It is mainly for technical and/or economical reasons in the majority of cases that monolithic augers made of hard metal or ceramic cannot be used. We therefore have to choose from augers made of hardened steels or high chrome cast steel, augers with welded surfaces ranging from PTA hard-facing, HVOF coatings through to ceramic coatings, PVD and CVD. Other augers are those with soldered or glued-on small plates of hard metal or ceramic, and in certain cases, augers made of plastic materials such as POM, as well as augers which have been kolsterized or manufactured from powder-metallurgical steels.

Apart from the consideration of whether the use of certain materials for augers and liners can be justified economically, certain restrictions must be accepted in handling bodies which are both abrasive and corrosive. There exist only very few stainless steels which can be hardened, as this always entails the risk that the corrosion resistance will suffer from this treatment. As far as coatings are concerned, care must be taken that the surfaces are not only wear-resistant but also, especially in case of augers, that the surfaces are as smooth as possible for as long as possible, are non-adhesive and have a certain coating thickness. Most of the many coating methods applied nowadays do not exhibit the combination of these features. Augers which are polished when new show an increased roughness over the service life—see Chap. 15. The layer thicknesses of other coatings, for example, PVD or CVD, are still too thin for the extrusion of abrasive ceramic bodies. Result: At the moment, we are faced with a trade-off situation between price, wear resistance and surface roughness and have to find case by case the appropriate compromise.

22.2.3 The Sensor Gap

Up to this day, when we learn about fabulous new sensors all the time, there are still no sensors available which can accurately measure online the moisture content of the extruded column or of the prepared body. All sensors presently available on the market, no matter on which principle they function—infrared light, microwave or X rays, etc.—are not suitable for measuring the moisture content online with an accuracy of plus/minus 0.1–0.2%. Now as before, we have to rely on offline measurements with the aid of proven moisture analyser instruments. Bad luck, because all the key rheological parameters, call it extrudability or plasticity or consistency depend on the moisture content of the body. Consequently, any fluctuations are noticed only when it is too late, when they manifest themselves as pulsating column speeds, variations in the pressure rate, etc. Even in extrusion plants where synthetic ceramic materials are being processed under strictly controlled conditions, fluctuations in the body happen again and again, although this should really be ruled out. Further gaps also exist in sensors suitable for recording the state of wear. The same applies to non-wearing sensors for measuring the pressure rate, for continuous density measurement, etc. On the other hand, some progress has also been made, in the form of sensors for contactless measurement of the column speed, sensors for online monitoring of the condition of gearboxes or sensors for level control. Nowadays, every extrusion plant should not be without these.

22.3 The Contribution of Artificial Intelligence

Since the publication of the first edition of “The Art of Ceramic Extrusion”, we have registered numerous developments in ceramic materials, novel shaping methods–particularly with additive manufacturing, new applications of ceramic components, and with regard to software, especially AI, although it is open to question whether software is the correct term for AI in view of the enormous volume of data and the deep learning potential. From today’s perspective, what opportunities are emerging for the extrusion of ceramic bodies? Certainly, AI is opening up opportunities in the development of tailored programs for the faster control of extrusion systems. The quality of the programs can also be optimized with the help of AI as it empowers the learning extruder to draw on past operating data, current operating data, set values, and throughput rates so as to respond swiftly and expediently to any variations in the body. AI is already proving a valuable aid in the development of design solutions and opening up new possibilities in the development of ceramic bodies from several components, with the aim of optimizing desired properties and reducing costs.

22.4 En Route to the “Smart Extrusion Concept” (SEC)

Continual efforts for improvement on the one hand in accordance with methods like Kaizen and, on the other, awareness of the theoretical and technological gaps and, finally, active efforts towards realizing these improvements to closing these gaps can be considered marks of a successful strategy for the future. It has to be added that there must be the awareness and sensitivity to the fact that all our efforts and interventions have to take place within a cross-linked and complex socio-economic system with all associated technical and social implications.

To end this short book, I would like to summarize seven important elements within the framework of the “Smart Extrusion Concept” (SEC) approach, in the hope that you will find these suggestions useful and will supplement them with your own elements.

- Jointly compile with your team during a brainstorming session a “long list” of problems and possibilities for improvement of the extrusion process. Let everyone come up with “crazy ideas” as long as these are of a concrete nature. Do not neglect those problems which have their causes outside extrusion, and do not forget to watch the market and possible, disruptive technologies and products.
- Generate a “long list” with all the ideas/suggestions, and try to evaluate and categorize these ideas in three categories:
  - High priority and realistic
  - Nice to have in the near future
  - Not yet available but highly desirable.

From this, compile a "short list" which you are able to work through and implement within a defined timeframe.

- Seek targeted information at regular intervals in respect of the latest developments and R&D findings by visiting relevant trade fairs, congresses, via Internet, pertinent literature and by talking to expert colleagues regarding the current state of the art and any research findings. Also take a look "over the fence" to see what happens in other related fields, such as extrusion of plastics, foodstuff or metals. Check the basis of your "long list", and ascertain whether new solutions are being offered to problems which still appeared to be insoluble at the time of compiling the list.
- Include in your considerations the opportunities offered by modern control and sensor technology, as well as the "Internet of things", in particular in respect of the monitoring of the extrusion system plus its subsystems and their interdependencies with other subsystems. Whether it makes sense to control the extrusion plant from your smartphone is of less importance than an answer to the question of which relevant operating parameters need to be controlled and be modified from time to time.
- Acquire more in-depth knowledge regarding the rheological properties of your ceramic bodies, ensure that the relevant characteristics, notably viscosity, yield point and wall friction are measured at regular intervals and determine the corridors within which the different extrudates should be produced.
- On the basis of the quantitative and qualitative product characteristics, check whether the extrusion plant is the optimum solution in its entirety. If the answer is "no", determine the source of the problems and how these can be solved.
- Do not rely on your own personal experience under the motto: "We tried that already some 20 years ago", but rather question that experience in the light of new boundary conditions, technical progress and the increased challenges.

Good Luck!

Bibliography

1. D. Kahneman, *Thinking* (Fast and Slow, Penguin, London, 2012)
2. H. Mintzberg, *Strategy Safari: The Complete Guide Through the Wilds of Strategic Management* (Pearson Education, New York, 2008)
3. T. Ohno, *Toyota Production System: Beyond Large-Scale Production* (Taylor & Francis, Oxford, 1988)

MIX
Papier aus verantwortungsvollen Quellen
Paper from responsible sources
FSC® C105338
FSC
www.fsc.org

If you have any concerns about our products,
you can contact us on
ProductSafety@springernature.com

In case Publisher is established outside the EU,
the EU authorized representative is:
Springer Nature Customer Service Center GmbH
Europaplatz 3, 69115 Heidelberg, Germany

Printed by Libri Plureos GmbH
in Hamburg, Germany